D1603467

SYSTEM IDENTIFICATION
Methods and Applications

APPLIED MATHEMATICS AND COMPUTATION

A Series of Graduate Textbooks, Monographs, Reference Works

Series Editor: ROBERT KALABA, University of Southern California

No. 1 MELVIN R. SCOTT
Invariant Imbedding and its Applications to Ordinary Differential Equations: An Introduction, 1973

No. 2 JOHN CASTI and ROBERT KALABA
Imbedding Methods in Applied Mathematics, 1973

No. 3 DONALD GREENSPAN
Discrete Models, 1973

No. 4 HARRIET H. KAGIWADA
System Identification: Methods and Applications, 1974

In Preparation:

No. 5 V. K. MURTHY
The General Point Process: Applications to Bioscience, Medical Research, and Engineering, 1974

SYSTEM IDENTIFICATION

Methods and Applications

Harriet H. Kagiwada
The Rand Corporation, Santa Monica, California

1974
Addison-Wesley Publishing Company
Advanced Book Program
Reading, Massachusetts

London · Amsterdam · Don Mills, Ontario · Sydney · Tokyo

CODEN: APMCC

Library of Congress Cataloging in Publication Data

Kagiwada, H 1937–
System identification: methods and applications.

(Applied mathematics and computation, no. 4)
Includes bibliographical references.
1. System analysis. 2. Inverse problems
(Differential equations) I. Title. [DNLM:
1. Mathematics. QA297 K12s 1973]
QA402.K29 003 73–19785
ISBN 0–201–04108–1
ISBN 0–201–04109–X (pbk.)

Reproduced by Addison-Wesley Publishing Company, Inc., Advanced Book Program, Reading, Massachusetts, from camera-ready copy prepared by the author.

American Mathematical Society (MOS) Subject Classification Scheme (1970):
93B30, 35R30, 34B15, 65L10

Manufactured in the United States of America

To Julia and Conan

CONTENTS

		Page
SERIES EDITOR'S FOREWORD		xv
PREFACE		xvii
1-1	Introduction	1
1-2	Determination of Potential	7
1-3	Quasilinearization, System Identification and Nonlinear Boundary Value Problems	11
1-4	Solution of the Potential Problem	18
	References	28
CHAPTER 2	IDENTIFICATION OF LAYERED MEDIA WITH MULTIPLE SCATTERING DATA	31
2-1	Introduction	31
2-2	Invariant Imbedding	32
2-3	The Diffuse Reflection Function for an Inhomogeneous Slab	34
2-4	Gaussian Quadrature	40
2-5	An Inverse Problem	43
2-6	Formulation as a Nonlinear Boundary Value Problem	48
2-7	Numerical Experiments I. Determination of c, the Thickness of the Lower Layer	49
2-8	Numerical Experiments II. Determination of T, the Overall Optical Thickness	54
2-9	Numerical Experiments III. Determination of the Two Albedos and the Thickness of the Lower Layer	55
2-10	Discussion	57
	References	58

CHAPTER 3 IDENTIFICATION USING NOISY SCATTERING MEASUREMENTS 61

3-1 Introduction 61
3-2 An Inverse Problem 62
3-3 Formulation as a Nonlinear Boundary Value Problem 63
3-4 Solution via Quasilinearization 65
3-5 Numerical Experiments I: Many Accurate Observations 70
3-6 Numerical Experiments II: Effect of Angle of Incidence 73
3-7 Numerical Experiments III: Effect of Noisy Observations 73
3-8 Numerical Experiments IV: Effect of Criterion 75
3-9 Numerical Experiments V: Construction of Model Atmospheres 77

References 82

CHAPTER 4 INVERSE PROBLEMS IN RADIATIVE TRANSFER: ANISOTROPIC SCATTERING 83

4-1 Introduction 83
4-2 The S Function 84
4-3 An Inverse Problem 91
4-4 Method of Solution 92
4-5 Numerical Results 96

References 98

CHAPTER 5 AN INVERSE PROBLEM IN NEUTRON TRANSPORT THEORY 101

5-1 Introduction 101
5-2 Formulation 102
5-3 Dynamic Programming 104
5-4 An Approximate Theory 106
5-5 A Further Reduction 110
5-6 Computational Procedure 111
5-7 Computational Results 112

References 122

CHAPTER 6 SYSTEM IDENTIFICATION BY MEASUREMENT OF TRANSIENT WAVES 123

6-1 Introduction 123
6-2 The Wave Equation 124
6-3 Laplace Transforms 125
6-4 Formulation 127
6-5 Solution via Quasilinearization 128
6-6 Example 1 - Homogeneous Medium, Step Function Force 131
6-7 Example 2 - Homogeneous Medium, Delta-Function Force 136
6-8 Example 3 - Inhomogeneous Medium with Delta-Function Input 139
6-9 Discussion 142

References 143

CHAPTER 7 STEADY STATE WAVE PROPAGATION. 145

7-1 Introduction 145
7-2 Some Fundamental Equations 147
7-3 Invariant Imbedding and the Reflection Coefficient 148
7-4 Production of Observations 152
7-5 Determination of Refractive Index 155
7-6 Numerical Experiments 159
7-7 Discussion 162

References 163

CHAPTER 8 PROBING AN INHOMOGENEOUS MEDIUM WITH RAYS 165

8-1 Introduction 165
8-2 Statement of the Problem 166
8-3 Quasilinearization 168
8-4 Methods of Solution of the Inverse Problem 171
8-5 Numerical Method and Results 174
8-6 Discussion 177

References 178

CHAPTER 9 NONLINEAR FILTERING OF SIGNALS 179

9-1 Introduction 179
9-2 Formulation 179
9-3 Invariant Imbedding 181
9-4 Continuation of the Analysis 183
9-5 Equation for the Weighting Factor 184
9-6 Practical Considerations 185
9-7 Numerical Results 186
9-8 System Identification 189
9-9 Numerical Results 190
9-10 Summary and Discussion 193

References 196

CHAPTER 10 NONLINEAR INTERPOLATING FILTER FOR IMPRECISE DYNAMICAL EQUATIONS 199

10-1 Introduction 199
10-2 Formulation of the Problem 200
10-3 A Two-Point Boundary Value Problem 202
10-4 The Equations of Invariant Imbedding 203
10-5 A Linear System 206
10-6 An Approximate Solution for the Nonlinear Problem 209
10-7 Conclusions 211
10-8 Vector Generalization 212

References 220

CHAPTER 11 DETERMINATION OF TIME LAGS 221

11-1 Introduction 221
11-2 Formulation 222
11-3 Method of Solution 224
11-4 Numerical Example 226
11-5 Discussion 228

References 229

CHAPTER 12 IDENTIFICATION PROBLEM OF CARDIOLOGY 231

12-1 Introduction 231
12-2 Basic Assumptions 233
12-3 The Inverse Problem 235
12-4 Quasilinearization 236
12-5 Numerical Experiments 237
12-6 Discussion 239

References 241

APPENDIX A PROBLEMS 243

APPENDIX B FORTRAN PROGRAMS 247

B-1 Remarks 247
B-2 Programs for Orbit Determination 247
Program A.1 Production of Observations 247
Program A.2 Determination of Orbit 249
B-3 Programs for Identification of Layered Media 256
Program B.1 Determination of c, the Thickness of the Interface 256
Program B.2 Determination of T, the Overall Optical Thickness 266
Program B.3 Determination of the Two Albedos and the Thickness of the Lower Layer 276

APPENDIX C LIBRARY ROUTINES 287

C-1 Library Routine INTS/INTM 287
C-2 Library Routine MATINV 288

INDEX 291

SERIES EDITOR'S FOREWORD

Execution times of modern digital computers are measured in nanoseconds. They can solve hundreds of simultaneous ordinary differential equations with speed and accuracy. But what does this immense capability imply with regard to solving the scientific, engineering, economic, and social problems confronting mankind? Clearly, much effort has to be expended in finding answers to that question.

In some fields, it is not yet possible to write mathematical equations which accurately describe processes of interest. Here, the computer may be used simply to simulate a process and, perhaps, to observe the efficacy of different control processes. In others, a mathematical description may be available, but the equations are frequently difficult to solve numerically. In such cases, the difficulties may be faced squarely and possibly overcome; alternatively, formulations may be sought which are more compatible with the inherent capabilities of computers. Mathematics itself nourishes and is nourished by such developments.

Each order of magnitude increase in speed and memory size of computers requires a reexamination of computational techniques and an assessment of the new problems which may be brought within the realm of solution. Volumes in this series will provide indications of current thinking regarding problem formulations, mathematical analysis, and computational treatment.

ROBERT KALABA

Los Angeles, California
April, 1973

PREFACE

System identification problems, or inverse problems, are found everywhere in science, medicine, and engineering. In such a problem, fundamental properties of a system are to be determined from observed behavior of that system. An example is that of determining the orbit and mass of a heavenly or man-made body.

A wide class of inverse problems may be numerically resolved by the use of modern mathematical and computational methods and high speed computers.

Let us first note that many direct problems may be formulated in terms of systems of ordinary differential equations of the form

$$\dot{x} = f(x,\alpha) \ . \tag{1}$$

Here, t is the independent variable, x is an n-dimensional vector whose components are the dependent variables, and α is an m-dimensional vector whose components represent the structure of the system. When the parameters in α and a complete set of initial conditions,

$$x(0) = c \ , \tag{2}$$

are known, a numerical integration of Eq. (1) produces the solution $x(t)$ on the interval $0 \leq t \leq T$. This initial value problem can be readily solved with a digital or an analog computer.

On the other hand, in an inverse or identification problem, the quantity $x(t)$ or some function of $x(t)$ is approximately known at various times, while the parameters are not directly determinable. It is desired to estimate the structure of the system as expressed by the parameter vector, α, and a complete set of initial conditions, c. This may be regarded as a nonlinear boundary value problem in which the unknowns are some or all of the c's and α's. There must be agreement with the observations, such as

$$x(t_i) \cong b_i, \qquad b_i = \text{given}. \tag{3}$$

Frequently, problems which do not naturally occur in the form of systems of ordinary differential equations may be expressed in that form in an approximate representation. In this book, it is shown how differential integral equations may be reduced to systems of ordinary differential equations with the use of a quadrature formula. Also, partial differential equations, such as the wave equation, may be expressed in the desired form by applying Laplace transform methods which remove the time derivative. Other possibilities are clearly open.

Recently, a comprehensive theory has been developed for the conversion of integral equations, boundary value problems, and variational problems into initial value problems. This means that the ideas presented in this book can be greatly extended over a wider class of inverse problems.

Nonlinear boundary value problems can be solved by a variety of methods among which are quasilinearization, dynamic programming, and invariant imbedding. These methods are especially suited to modern computers, for they reduce nonlinear boundary value problems to nonlinear initial value problems which are more readily treated on electronic computers.

The foregoing ideas are illustrated in this book by the formulation and solution of some inverse problems which arise in celestial mechanics, radiative transfer, neutron transport, wave propagation, geometrical optics, nonlinear filtering, and physiology. Numerical experiments are conducted to estimate the accuracy and stability of the methods, and the effect of number and quality of observational measurements. Exercises and selected FORTRAN programs are given in the Appendices.

These concepts may be helpful in the planning of experiments and in the choice of apparatus. They may be applied to the designing of systems to have certain desired properties, and will be useful in the construction of mathematical models. It is hoped that these studies in system identification and inverse problems will prove to be of value to many other doctors, scientists and engineers as we know they have been to some.

This book is the outgrowth of research performed at The Rand Corporation, Kyoto University, and elsewhere, and a part of it has been included in the author's doctoral dissertation. She wishes to express her appreciation to her dear friends and colleagues, Sueo Ueno and Robert Kalaba, as well as to many others who have contributed to the making of this book. Special thanks go to Lynn Anderson and Sandy Whitaker for their expert typing and cheerful assistance in the preparation of this book.

Harriet H. Kagiwada

SYSTEM IDENTIFICATION
Methods and Applications

CHAPTER 1

INTRODUCTION

1-1 INTRODUCTION

Inverse problems, or system identification problems, are fundamental problems of science, medicine and engineering [1-12]. Man has always sought knowledge of a physical system beyond that which is directly observable. Even today, we try to understand the dynamical processes of the deep interior of the sun by observing the radiation emerging from the sun's surface. We deduce the potential field of an atom from nuclear scattering experiments. We attempt to determine the state of a person's heart by reading electrocardiograms. The underlying theme is the relationship between the internal structure of a system and the observed output. The hidden features of the system are to be extracted from the experimental data.

Mathematical treatment of physical problems has been devoted almost exclusively to the "direct problem." A complete picture of the system is assumed to be given, and equations are derived which describe the output as a function of the system parameters. The inverse problem is to determine the parameters and structure of a system as a function of the observed output.

It is possible to solve a given inverse problem by solving a series of direct problems: by assuming different sets of parameters, determining the corresponding outputs from the theoretical equations, and comparing theoretical *versus* experimental results. By trial and error, one may find a solution which approximately agrees with the experimental data. This is not a very efficient procedure. Another way to proceed is to solve analytically for the unknown parameters as functions of the measurements. This method generally requires much abstract mathematics and simple approximations of complex functions. The resultant inverse solution may be valid only in very special circumstances.

What we seek are efficient, systematic procedures for solving a wide class of system identification and inverse problems - procedures which are suitable for execution on high speed electronic computers. Computers are currently capable of integrating large systems of ordinary differential equations, given a complete set of initial conditions, with high accuracy. We would therefore like to formulate our problems in terms of systems of ordinary differential equations. Partial differential equations, such as the wave equation,

$$\frac{\partial^2 u(x,t)}{\partial x^2} = \frac{1}{c^2} \frac{\partial^2 u(x,t)}{\partial t^2} \tag{1.1}$$

may be reduced to systems of ordinary differential equations in several ways. These include the use of Laplace transform methods, Fourier decomposition, and finite difference schemes. Integro-differential equations, which frequently occur in transport theory, may be reduced to systems of ordinary differential equations by approximating the definite integrals by finite sums using Gaussian and other quadrature formulas. Other means of formulating problems in terms of ordinary differential equations are possible.

An exact theory has recently been developed for the reduction of integral equations, boundary-value problems, and variational problems into initial value problems [13-18]. We can therefore expect greater extension and applicability of the methods presented here for system identification.

We desire to formulate our identification problem in such a way that we deal with ordinary differential equations. First, as we shall show, we may express the problem as a nonlinear boundary value problem, in which we seek a complete set of initial conditions. The unknown system parameters will be calculated directly from the initial conditions. Next, we resolve the nonlinear boundary value problem, ordinarily a difficult task, by the use of some sophisticated techniques [19-28]. We may replace the nonlinear boundary value problem by a rapidly converging sequence of linear boundary value problems via the technique of quasilinearization [11, 21, 22]. We may, alternatively, treat the problem as a multi-stage decision process with the use of dynamic programming [23]. Or, we may solve directly for the missing initial conditions by applying the concept of invariant imbedding [19, 28]. From the solution of the nonlinear boundary value problem, we immediately obtain knowledge of the internal structure of the system.

In this book, we discuss some of these relatively new concepts, computational techniques, and applications. Our examples from the physical, biological and engineering sciences are physically motivated. No specialized background is required on the part of the reader. We intend to be self-contained in the mathematical derivations, except for those matters which are well-treated elsewhere, such as dynamic programming, linear programming, and the numerical inversion of Laplace transforms. Again, no special mathematical knowledge is needed beyond the level of ordinary differential equations and linear algebraic equations. We will, however, assume that we have at our disposal a high-speed digital computer with a memory of about 32,000 words, plus a library of computer routines for numerical integration, matrix inversion, and linear programming. Our basic assumption is that our computer can integrate large systems of ordinary differential equations rapidly and accurately [29, 30].

In the first chapter, we wish to emphasize some important ideas. We are given geocentric observations of a heavenly body, taken at various times [31-32]. The orbit of this body lies in the potential field created by the sun and an unknown perturbing mass. We show how the mass may be identified and the orbital elements found. For simplicity, we assume that the position of the perturbing mass is given; if desired, the position as a function of time could also be estimated. Since we are virtually forced by our modern computers to take a fresh look at old problems, we are not concerned with conic sections. A new methodology, based on high speed digital computers, is developed. The technique of quasilinearization, described in this chapter, enables us to solve this inverse problem with a minimum of effort. In spite of the newness of this solution of a long-standing problem in celestial mechanics, we employ this example for purely illustrative purposes.

Transport theory is intimately concerned with the determination of radiation fields within scattering and absorbing media. Our first problem in radiative transfer (Chapter 2) serves to exemplify the philosophy and application of invariant imbedding. We derive the basic integro-differential equation for the diffuse reflection function, and we reduce it to a system of ordinary differential equations by the method of Gaussian quadrature. Then we formulate an inverse problem for the determination of layers in a medium from knowledge of the diffusely reflected light. We outline the computational procedure, and we present our results. In Chapter 3, our setting is again an inhomogeneous scattering medium. We investigate the effects of errors in our measurements, the number and quality of the observations, and the criterion function, on the estimates of the medium. Our criteria are either of least squares type, which leads to linear algebraic equations, or of minimax form, which is suitable for linear programming. We also consider a variation of the inverse problem, the construction of a model atmosphere according to certain specifications. In Chapter 4, we consider an anisotropically scattering medium. The phase function is to be determined on the basis of measurements of diffusely reflected radiation in various directions.

An inverse problem in neutron transport (Chapter 5) is solved in a novel way. The dynamic programming approach leads to a determination of absorption coefficients in a rod, from measurements of internal fields. The calculation is done by an exact method, and is compared with a calculation based on an approximate theory. The approximate theory is accurate and less costly in computing time.

As we have already mentioned, the partial differential wave equation may be reduced to a system of ordinary differential equations by Laplace transform methods or by Fourier decompositions. In Chapter 6, we deal with ordinary differential equations for the Laplace transforms of the disturbances. In these equations, time appears only as a parameter. Our measurements of the disturbances at various times are converted to the corresponding transforms by means of Gaussian quadrature. We solve a nonlinear boundary value problem in order to determine the system parameters. The inverse Laplace transforms may be obtained by a numerical inversion technique [26].

In Chapter 7, we use a decomposition of the form $u(x,t) = u(x)e^{-i\omega t}$, corresponding to a steady-state situation of wave propagation. We probe an inhomogeneous slab with waves of different frequencies and we "measure" the reflection coefficients. We wish to determine the index of refraction as a function of distance in the medium. Invariant imbedding leads to ordinary differential equations for the reflection coefficients, with known initial conditions. The unknown index of refraction in the equations and the observations of terminal values of the reflection coefficients make this a nonlinear boundary value problem. Quasilinearization is used to solve the problem, and computational results are presented.

In Chapter 8, we estimate the index of refraction of an inhomogeneous medium as a function of two spatial coordinates. This is accomplished by considering the arrival times of various rays.

Observations of systems undergoing nonlinear processes are converted into sequential estimates in Chapter 9, when the equations are assumed to be exact, and later extended in Chapter 10 to include inexact knowledge of the dynamical equation.

Differential difference equations are also reducible to ordinary differential equations, as we demonstrate in Chapter 11, and we further proceed to determine unknown time lags.

The inverse problem of cardiology, i.e., the determination of the electrical state of the heart based on vectorcardiograms, is discussed and numerical results are presented in Chapter 12.

1-2 DETERMINATION OF POTENTIAL

Consider the motion of a particle (or a wave) in a potential field $V = V(x, y, z; k_1, k_2, \ldots, k_n)$ where we recognize the dependence on physical parameters $k_1, k_2, \ldots, k_n$. Suppose that these parameters are unknown, and that we have observations of the motion of the particle at various times. We wish to determine the potential function on the basis of these measurements.

Consider the following situation. A heavenly body H of mass m moves in the potential field created by the sun and a perturbing body P, whose masses are M and m_p, respectively, and $m << m_p << M$. All of the bodies concerned lie in the ecliptic plane. The potential energy varies inversely as the distance from the sun, r_s, and from the perturbing body, r_p,

$$V = -\frac{k_s}{r_s} - \frac{k_p}{r_p} . \tag{1.2}$$

Here, k_s and k_p are the parameters

$$k_s = \gamma\, m\, M,\; k_p = \gamma\, m\, m_p, \tag{1.3}$$

where γ is the constant of gravitation. The quantity k_s may be assumed to be known. We choose our units so that $k_s \equiv m$, or $\gamma M \equiv 1$. The parameter k_p is unknown and $k_p < k_s$. We wish to determine k_p and thus V by observing the motion of H.

Let us take the plane of the ecliptic to be the (x, y) plane. The sun is situated at the origin, the earth at the point $(1, 0)$, and the perturbing body at the location $(\xi, \eta) = (4, 1)$. The earth only enters into the discussion as the point from which measurements are taken. Its mass is neglected. The potential function is

$$V(x, y; k_p) = -\frac{k_s}{(x^2 + y^2)^{1/2}} - \frac{k_s}{[(\xi - x)^2 + (\eta - y)^2]^{1/2}} . \tag{1.4}$$

Angular observations of H are made at various times t_i, $i = 1, 2, \ldots, 5$. Fig. 1-1 illustrates the physical situation. Each solid arrow points to H at a given time t_i. The angle between the line of sight and the x axis is the observation. For comparison, see the dashed arrows which point to H when the mass of P is exactly zero, i.e., when $k_p = 0$. It is obvious that k_p is small.

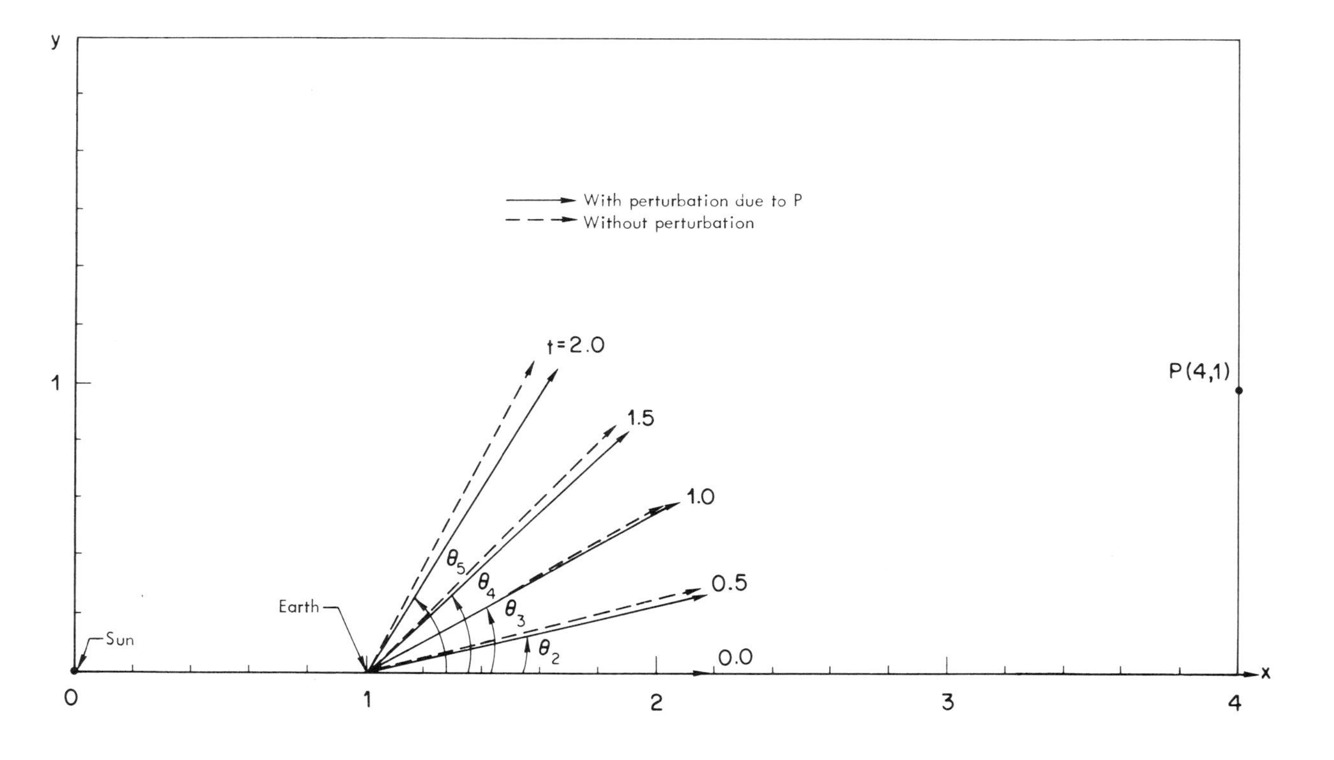

Figure 1-1 Angular Observations of a Heavenly Body

The equations of motion are

$$\ddot{x} = \frac{-x}{(x^2 + y^2)^{3/2}} + \frac{\alpha(\xi - x)}{[(\xi - x)^2 + (\eta - y)^2]^{3/2}},$$
$$\ddot{y} = \frac{-y}{(x^2 + y^2)^{3/2}} + \frac{\alpha(\eta - y)}{[(\xi - x)^2 + (\eta - y)^2]^{3/2}}, \tag{1.5}$$

where the parameter α,

$$\alpha = \frac{k_p}{k_s} = \frac{m_p}{M}, \tag{1.6}$$

is the mass of P relative to the mass of the sun. At times t_i, we obtain the angular data $\theta(t_i)$ which are, in radians,

$$\begin{aligned} \theta(0.0) &= 0.0, \\ \theta(0.5) &= 0.252188, \\ \theta(1.0) &= 0.507584, \\ \theta(1.5) &= 0.763641, \\ \theta(2.0) &= 1.01929. \end{aligned} \tag{1.7}$$

We wish to determine α, $x(0)$, $\dot{x}(0)$, $y(0)$, $\dot{y}(0)$ so that the conditions

$$\tan \theta(t_i) = \frac{-y(t_i)}{1 - x(t_i)} \tag{1.8}$$

are fulfilled. This is a nonlinear multipoint boundary value problem. The solution of this problem gives the relative mass of the perturbing body and the orbit of H as a function of time. The potential (1.4) is determined when α is known. We may consider the problem then to be the determination of the orbit [19, 24, 27, 31, 32].

For an arbitrary potential field, we are unable to express the solution analytically. We solve the problem computationally using the technique of quasilinearization[21, 22].

1-3 QUASILINEARIZATION, SYSTEM IDENTIFICATION AND NONLINEAR BOUNDARY VALUE PROBLEMS

Consider a physical system or process which is described by the system of N equations

$$\dot{x} = f(x, \alpha), \tag{1.9}$$

where x is a vector of dimension N, a function of independent variable t, with the N initial conditions

$$x(0) = c. \tag{1.10}$$

The vector x describes the state of the system at "time" t, and α is a parameter vector of the system. With α given, Eqs. (1.9) and (1.10) completely describe the system, for the state at any time t, $x(t)$, may be calculated by a numerical integration of (1.9) with initial conditions (1.10).

Now let us suppose that we have a system described by Eqs. (1.9), but α is unknown to us, and the initial conditions (1.10) are also unknown. However, we are able to make measurements of certain components of the state of the system at various times t_i. We wish to identify the system by determining α, and we wish to find a complete set of initial conditions $x(0) = c$ so that the system is fully described.

We think of the system parameter vector as if it were a dependent variable which satisfies the vector equation

$$\dot{\alpha} = 0, \tag{1.11}$$

with the unknown initial conditions

$$\alpha(0) = \alpha_0. \tag{1.12}$$

The multipoint boundary value problem which we have before us is to find the complete set of initial conditions

$$\begin{aligned} x(0) &= c, \\ \alpha(0) &= \alpha_0, \end{aligned} \tag{1.13}$$

such that the solution of the nonlinear system

$$\begin{aligned} \dot{x} &= f(x,\alpha), \\ \dot{\alpha} &= 0, \end{aligned} \tag{1.14}$$

agrees with the boundary conditions

$$x(t_i) = b_i, \tag{1.15}$$

where b_i is the observed state of the system at time t_i. Let us suppose that we have exactly $R = N + M$ measurements. of the first component of x, where N is the dimension of x and M is the dimension of α.

The boundary conditions are readily modified for a two point boundary value problem, or for more than R observations, or for other types of measurements, for example linear combinations of the components of x.

Our approach to the problem is one of successive approximations. We solve a sequence of linear problems. We assume only that large systems of ordinary differential equations, whether linear or nonlinear, may be accurately integrated numerically if initial conditions are prescribed, and that linear algebraic systems may be accurately resolved.

Let us define a new column vector x of dimension R, having as its elements the components of the original vector x and the components of α,

$$x = \begin{bmatrix} x_1 \\ x_2 \\ \\ \\ \\ \cdot \\ \cdot \\ \cdot \\ \\ \\ \\ x_R \end{bmatrix} = \begin{bmatrix} x_1 \\ x_2 \\ \cdot \\ \cdot \\ \cdot \\ x_N \\ \alpha_1 \\ \alpha_2 \\ \cdot \\ \cdot \\ \cdot \\ \alpha_M \end{bmatrix} . \tag{1.16}$$

This vector of dependent variables $x(t)$ satisfies the system of nonlinear equations

$$\dot{x} = f(x) , \tag{1.17}$$

according to (1.14), and it has the unknown initial conditions

$$x(0) = c, \tag{1.18}$$

according to (1.13). The boundary conditions are

$$x_1(t_i) = b_i, \quad i = 1, 2, \ldots, R. \tag{1.19}$$

Mathematically, we need not distinguish between the components of this new vector x as state variables or system parameters.

An initial approximation starts the calculations. We form an estimate of the initial vector c, and we integrate system (1.17) to produce the solution $x(t)$ over the time interval of interest, $0 \leqq t \leqq T$, via numerical integration. The quasilinearization procedure is applied iteratively until a convergence to a solution occurs, or the solution diverges.

Let us suppose that we have completed stage k of our calculations and we have the current approximation $x^k(t)$. In stage $k + 1$, we wish to calculate a new approximation $x^{k+1}(t)$.

The vector function $x^{k+1}(t)$ is the solution of the <u>linear</u> system

$$\dot{x}^{k+1} = f(x^k) + J(x^k)\,(x^{k+1} - x^k)\,, \tag{1.20}$$

where $J(x)$ is the Jacobian matrix with elements

$$J_{ij} = \frac{\partial f_i}{\partial x_j}\,. \tag{1.21}$$

Since x^{k+1} is a solution of a system of linear differential equations, we know from general theory that it may be represented as the sum of a particular solution, $p(t)$, and a linear combination of R independent solutions of the homogeneous equations, $h^i(t)$, $i = 1, 2, \ldots, R$,

$$x^{k+1}(t) = p(t) + \sum_{i=1}^{R} c^i\, h^i(t)\,. \tag{1.22}$$

The function p satisfies the equation

$$\dot{p} = f(x^k) + J(x^k)\,(p - x^k)\,, \tag{1.23}$$

and for convenience we choose the initial conditions

$$p(0) = 0\,. \tag{1.24}$$

The functions h^i are solutions of the homogeneous systems

$$\dot{h}^i = J(x^k)\, h^i\,, \tag{1.25}$$

and we choose the initial conditions

$$h^i(0) = \text{the unit vector with all of its components zero, except for the } i^{th} \text{ which is one.} \tag{1.26}$$

The $h^i(0)$ form a linearly independent set. If the interval $(0, T)$ is sufficiently small, the functions $h^i(t)$ are also independent. The solutions $p(t)$, $h^i(t)$ are produced by numerical integration with the given initial conditions. There are $R + 1$ systems of differential equations, each with R equations, making a total of $R(R + 1)$ equations which are integrated at each stage of our calculations.

After the functions p and h^i have been found over the interval, we must combine them so as to satisfy the boundary conditions (1.19),

$$b_i = p_1(t_i) + \sum_{j=1}^{R} c^j h_1^j(t_i), \quad i = 1, 2, \ldots, R. \tag{1.27}$$

This results in a system of R linear algebraic equations for the determination of the R unknown multipliers c^j, of the standard form

$$A\,c = B, \tag{1.28}$$

where the elements of the $R \times R$ matrix of coefficients A are

$$A_{ij} = h_1^j(t_i), \tag{1.29}$$

and the components of the R-dimensional column vector B are

$$B_i = b_i - p_1(t_i). \tag{1.30}$$

Having determined the multipliers, we now know a complete set of initial conditions for the $(k + 1)^{st}$ stage,

$$c = x^{k+1}(0) = p(0) + \sum_{j=1}^{R} c^j h^j(0) . \tag{1.31}$$

Because of our choice of initial conditions for p and h^j, the initial values for each component of the vector x are identical with the multipliers c^j,

$$c_i = x_i^{k+1}(0) = c^i, \quad i = 1, 2, \ldots, R . \tag{1.32}$$

Furthermore, we have a new approximation to the system parameter vector α,

$$\alpha_i = c^{N+i}, \quad i = 1, 2, \ldots, M. \tag{1.33}$$

The new approximation $x^{k+1}(t)$ for the interval $(0, T)$ may be produced either by the integration of the linear equations with the initial conditions just found, or by the linear combination of $p(t)$ and $h(t)$. The $(k + 1)^{st}$ cycle is complete and we are ready for the $(k + 2)^{nd}$. The process may be repeated until no further change is noted in the vector c.

The quasilinearization procedure is analogous to Newton's method for finding roots of an equation, $f(x) = 0$. If x^0 is an approximate value of one of the roots of $f(x)$, then an improved value x^1 is obtained by applying the Taylor expansion formula to $f(x)$, and neglecting higher derivatives,

$$f(x^1) = f(x^0) + (x^1 - x^0) \frac{\partial f(x^0)}{\partial x^0} . \tag{1.34}$$

Thus, the next approximation of the root is

$$x^1 = x^0 - \frac{f(x^0)}{f'(x^0)} . \tag{1.35}$$

In quasilinearization, if the function $x^0(t)$ is an approximate solution of the nonlinear differential equation,

$$\dot{x} = f(x), \tag{1.36}$$

then an improved solution $x^1(t)$ may be obtained in the following manner. The function $f(x)$ is expanded around the current estimate $x^0(t)$, neglecting higher derivatives,

$$f(x^1) = f(x^0) + (x^1 - x^0) \frac{\partial f(x^0)}{\partial x^0} . \tag{1.37}$$

The improved approximation $x^1(t)$ is the solution of the linear equation,

$$\dot{x}^1 = f(x^0) + (x^1 - x^0) \frac{\partial f(x^0)}{\partial x^0} . \tag{1.38}$$

The method is easily extended to vector functions, as we have seen. The sequence of functions $x^1(t)$, $x^2(t)$, $x^3(t)$, ..., may be shown to converge quadratically in the limit [22]. Practically speaking, a good initial approximation leads to rapid convergence, with the number of correct digits approximately doubling with each additional iteration. On the other hand, a poor initial approximation may lead to divergence.

The quasilinearization technique provides a systematic way of treating nonlinear boundary value problems. The computational solution of such a problem is broken up into stages, in which a large system of ordinary differential equations is integrated with known initial conditions, and a linear algebraic system is resolved. The initial value integration problem is well-suited to the digital computer. With the aid of a formula such as the trapezoidal rule,

$$\int_{t_0}^{t_n} f(t)dt \cong \frac{\Delta}{2}(f_0 + f_1) + \frac{\Delta}{2}(f_1 + f_2) + \dots + \frac{\Delta}{2}(f_{n-1} + f_n), \quad (1.39)$$

the integral of a function over an interval is rapidly computed. Moreover, higher order methods such as the Runge-Kutta and the Adams-Moulton, usually of fourth order, make it possible to solve the integration problem accurately and rapidly. The accuracy may be as high as one part in 10^8. The solution is available at each grid point t_0, $t_0 + \Delta$, $t_0 + 2\Delta$, ..., t_n, and may be stored in the computer's memory for use at some future time. The numerical integration of several hundred first order equations is a routine affair, barring instability.

On the other hand, the solution of a linear algebraic system is not a routine matter, computationally speaking. While formulas exist for the numerical inversion of a matrix, the solution may be inaccurate. The matrix may be ill-conditioned, and other techniques may have to be brought into play to remedy the situation. The storage of the nth approximation for the calculation of the $(n + 1)^{st}$ approximation may become a problem; a suggestion for overcoming this difficulty is given in [25].

1-4 SOLUTION OF THE POTENTIAL PROBLEM

We follow the method of quasilinearization to identify the unknown mass and to solve the problem of potential determination of Section 1-2. The nonlinear system of equations is

$$\begin{aligned}
\ddot{x} &= -\frac{x}{r^3} - \alpha\,\frac{x-\xi}{s^3}\,, \\
\ddot{y} &= -\frac{y}{r^3} - \alpha\,\frac{y-\eta}{s^3}\,, \\
\dot{\alpha} &= 0
\end{aligned} \tag{1.40}$$

with

$$r^2 = x^2 + y^2,\; s^2 = (x-\xi)^2 + (y-\eta)^2 \,. \tag{1.41}$$

Eqs. (1.40) are equivalent to a system of five first order equations for x, $\dot{x}$, y, $\dot{y}$, and α. The system of linear equations for the $(k+1)^{st}$ stage is

$$\begin{aligned}
\ddot{x}^{k+1} &= \left\{ -\frac{x^k}{r^3} - \alpha^k\,\frac{x^k-\xi}{s^3} \right\} \\
&+ (x^{k+1} - x^k)\left\{ -\frac{1}{r^3} + \frac{3x^{k2}}{r^5} - \frac{\alpha^k}{s^3} + \frac{3\alpha^k(x^k-\xi)^2}{s^5} \right\} \\
&+ (y^{k+1} - y^k)\left\{ \frac{3x^k y^k}{r^5} + \frac{3\alpha^k(x^k-\xi)(y^k-\eta)}{s^5} \right\} \\
&+ (\alpha^{k+1} - \alpha^k)\left\{ -\frac{x^k-\xi}{s^3} \right\},
\end{aligned} \tag{1.42}$$

$$\begin{aligned}
\ddot{y}^{k+1} &= \left\{ -\frac{y^k}{r^3} - {}^k\,\frac{x^k-\xi}{s^3} \right\} \\
&+ (x^{k+1} - x^k)\left\{ -\frac{3x^k y^k}{r\;r^5} + \frac{3\alpha^k(x^k-\xi)(y^k-\eta)}{s^5} \right\} \\
&+ (y^{k+1} - y^k)\left\{ -\frac{1}{r^3} + \frac{3y^{k2}}{r^5} - \frac{\alpha^k}{s^3} + \frac{3\alpha^k(y^k-\eta)^2}{s^5} \right\} +
\end{aligned}$$

$$+ (\alpha^{k+1} - \alpha^{k}) \quad - \frac{y^{k}-\eta}{s^3} \ , \tag{1.42}$$

$$\dot{\alpha}^{k+1} = 0 \ ,$$

where

$$r^2 = (x^k)^2 + (y^k)^2, \ s^2 = (x^k - \xi)^2 + (y^k - \eta)^2 \ . \tag{1.43}$$

We express the solution of (1.42) as the sum of a particular solution of (1.42) plus a linear combination of five independent solutions of the homogeneous form of (1.42),

$$x^{k+1}(t) = p_x(t) + \sum_{j=1}^{5} c^j \, h_x^j(t) \ ,$$

$$y^{k+1}(t) = p_y(t) + \sum_{j=1}^{5} c^j \, h_y^j(t) \ , \tag{1.44}$$

$$\alpha^{k+1}(t) = p_\alpha(t) + \sum_{j=1}^{5} c^j \, h_\alpha^j(t) \ .$$

Here, the symbol $p_x(t)$ is meant to represent the x component of the particular solution, which is a vector of dimension five, and similarly for the symbols $p_y(t)$, $p_\alpha(t)$. The symbols $h_x^j(t)$, $h_y^j(t)$, $h_\alpha^j(t)$ respectively correspond to the x, y, and α components of the jth homogeneous solution vector, for $j=1, 2, \ldots, 5$. The system which the particular solution satisfies is

$$\ddot{p}_x = -\{\frac{x^k}{r^3} - \alpha^k \frac{x^k - \xi}{s^3}\}$$

$$+ (p_x - x^k)\ \{-\frac{1}{r^3} + \frac{3x^{k^2}}{r^5} - \frac{\alpha^k}{s^3} + \frac{3\alpha^k (x^k - \xi)^2}{s^5}\}$$

$$+ (p_y - y^k)\ \{\frac{3x^k y^k}{r^5} + \frac{3\alpha^k (x^k - \xi)(y^k - \eta)}{s^5}\}$$

$$+ (p_\alpha - \alpha^k)\ \{-\frac{x^k - \xi}{s^3}\},$$

$$\ddot{p}_y = \{-\frac{y^k}{r^3} - \alpha^k \frac{y^k - \eta}{s^3}\} \tag{1.45}$$

$$+ (p_x - x^k)\ \{\frac{3x^k y^k}{r^5} + \frac{3\alpha^k (x^k - \xi)(y^k - \eta)}{s^5}\}$$

$$+ (p_y - y^k)\ \{-\frac{1}{r^3} + \frac{3y^{k^2}}{r^5} - \frac{\alpha^k}{s^3} +, \frac{3\alpha^k (y^k - \eta)^2}{s^5}\}$$

$$+ (p_\alpha - \alpha^k)\ \{-\frac{y^k - \eta}{s^3}\},$$

$$\dot{p}_\alpha = 0,$$

with the initial condition

$$p(0) = 0. \tag{1.46}$$

The jth homogeneous solution satisfies the system

$$\ddot{h}^j_x = h^j_x\ \{-\frac{1}{r^3} + \frac{3x^{k^2}}{r^5} - \frac{\alpha^k}{s^3} + \frac{3\alpha^k (x^k - \xi)^2}{s^5}\}$$

$$+ h^j_y\ \{\frac{3x^k y^k}{r^5} + \frac{3\alpha^k (x^k - \xi)(y^k - \eta)}{s^5}\} \quad + \tag{1.47}$$

$$+ h_\alpha^j \{- \frac{x^k - \xi}{s^3}\} ,$$

$$\ddot{h}_y^j = h_x^j \frac{3x^k y^k}{r^5} + \frac{3\alpha^k (x^k - \xi)(y^k - \eta)}{s^5}\}$$

$$+ h_y^j \{- \frac{1}{r^3} + \frac{3y^{k^2}}{r^5} - \frac{\alpha^k}{s^3} + \frac{3\alpha^k (y^k - \eta)^2}{s^5}\} \quad (1.47)$$

$$+ h_\alpha^j \{- \frac{y^k - \eta}{s^3}\},$$

$$\dot{h}_\alpha^j = 0.$$

Its five initial conditions are presented in the appropriate column of Table 1-1.

Table 1-1

The Initial Conditions for the Homogeneous Solutions

	j=1	2	3	4	5
$h_x^j(0)$	1	0	0	0	0
$\dot{h}_x^j(0)$	0	1	0	0	0
$h_y^j(0)$	0	0	1	0	0
$\dot{h}_y^j(0)$	0	0	0	1	0
$h_\alpha^j(0)$	0	0	0	0	1

The particular and homogeneous solutions are produced by numerical integration and are known at the discrete times $t = 0, \Delta, 2\Delta, 3\Delta, \ldots, T$.

Let us find the system of linear algebraic equations which is to be solved in the $(k+1)^{st}$ stage. The boundary conditions may be expressed as

$$y^{k+1}(t_i) + [1-x^{k+1}(t_i)] \tan \theta(t_i) = 0, \tag{1.48}$$

where $\theta(t_i)$ is the observed angular position of the heavenly body H at time t_i. Using relations (1.44), we obtain the five equations

$$\sum_{j=1}^{5} c^j \; [h_y^j(t_i) - h_x^j(t_i) \tan \theta(t_i)]$$

$$= - \tan \theta(t_i) - p_y(t_i) + p_x(t_i) \tan \theta(t_i), \tag{1.49}$$

$$i = 1, 2, \ldots, 5,$$

for the five unknowns $c^1, c^2, \ldots c^5$.

The solution of (1.49) immediately gives us our new set of orbital parameters and the mass of the unknown perturbing body P,

$$\begin{aligned} x^{k+1}(0) &= c^1, \\ \dot{x}^{k+1}(0) &= c^2, \\ y^{k+1}(0) &= c^3, \end{aligned} \tag{1.50}$$

$$\dot{y}^{k+1}(0) = c^4,$$

$$\alpha^{k+1}(0) = c^5.$$

Since we need $x^{k+1}(t)$, $y^{k+1}(t)$, and $\alpha^{k+1}(t)$, for stage $k+2$, we use relations (1.44) for the evaluation of these functions at $t = 0, \Delta, 2\Delta, 3\Delta, \ldots, T$. The cycle is ready to begin once more, and it is repeated until a solution of the nonlinear problem is found, or for a fixed number of stages.

We begin a numerical experiment with the initial guess that at time $t = 0$, the body H is at location (3,0) with velocity coordinates $x = 0$, $y = 1$, and we believe that the mass of P is about 0.3. We integrate equations (1.40) with the initial values

$$x(0) = 3,\ \dot{x}(0) = 0,\ y(0) = 0,\ \dot{y}(0) = 1,\ \alpha(0) = 0.3\ , \tag{1.51}$$

from $t = 0$ to $t = 2.5$, using a grid size of $\Delta = 0.01$ and an Adams-Moulton integration formula. This generates the curve labelled "Initial Approximation" in Fig. 1-2. This is a very poor approximation to the true orbit. After two stages of the quasilinearlization scheme, our approximation has improved so that the orbit is represented by the curve labelled "Approximation 2" in Fig. 1-2. In five iterations, we converge to the true curve, $h(x,y)$, and we have found the correct value of 0.2 for the mass of the perturbing body. The rate of convergence is indicated in Table 1-2.

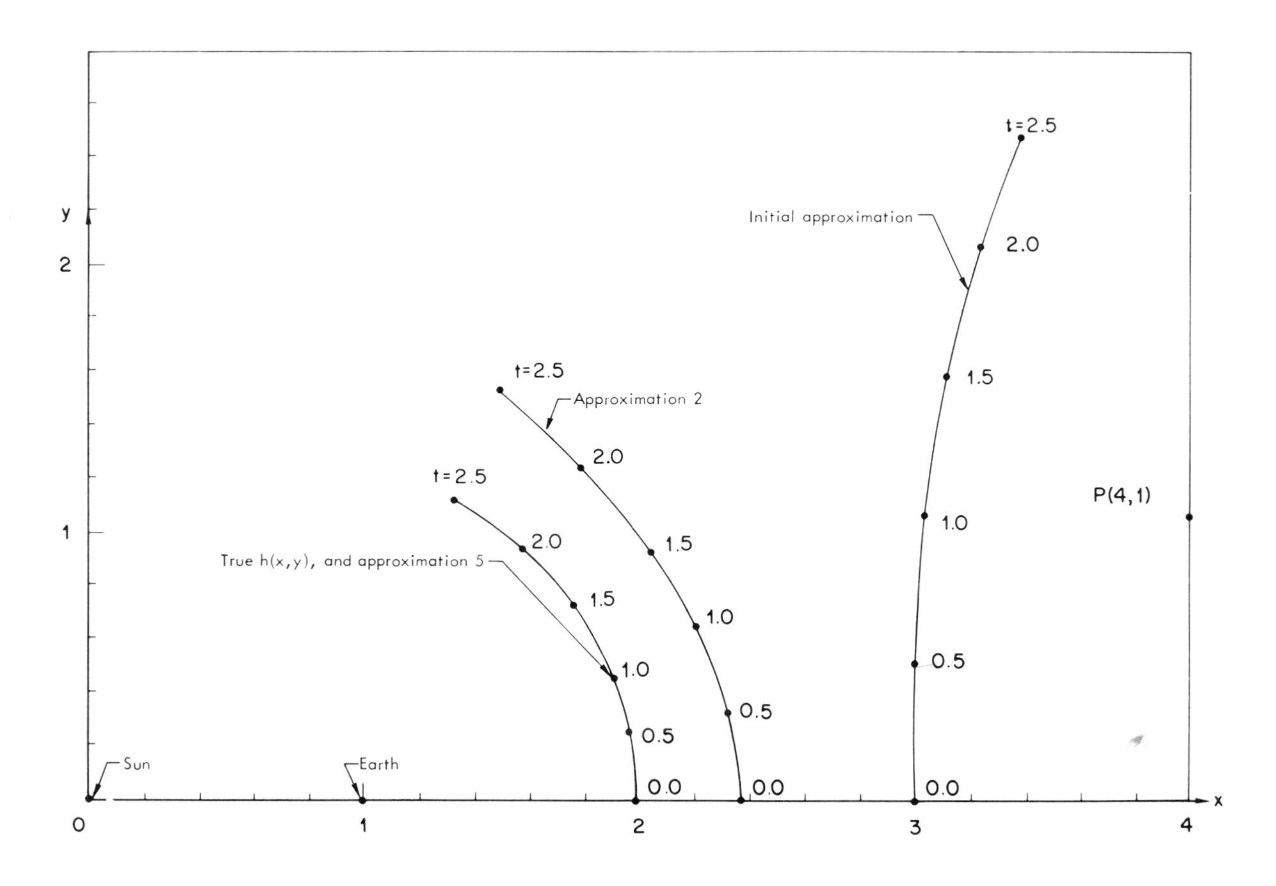

Figure 1-2 Successive Approximations of the Orbit

Table 1-2

Successive Approximations of the Complete Set of Initial Conditions and the Mass of P

Approx.	$x(0)$	$\dot{x}(0)$	$y(0)$	$\dot{y}(0)$	α
0	3.0	0.0	0.0	1.0	0.3
1	3.18421	-.221272	0.0	1.06544	-.120164
2	2.37728	-.061370	0.0	0.690767	-.259144
3	2.11189	-.018545	0.0	0.555666	-.070333
4	2.01974	-.003194	0.0	0.509813	.141922
5	2.00023	.000013	0.0	0.500120	.198208
True	2.0	0.0	0.0	0.5	0.2

Suppose that we also wish to know the position of H at some "future" time $t = 2.5$. Our sequence of approximations of the predicted location is given in Table 1-3. The entire calculations require only 1-1/2 minutes on the IBM 7044 computer, using a FORTRAN IV source language. The FORTRAN programs which generate the data and which determine the orbit and mass are listed in Appendix B.

The time involved is mainly due to the evaluation of the derivatives of the functions. The Adams-Moulton fourth order method requires the derivative to be evaluated twice for each integration step forward [29].

In another trial, beginning with the approximation that the orbit is a point at the earth's center, we find another solution which satisfies all of the conditions. However, the mass turns out to be greater than one, an unallowed solution.

Table 1-3

Predicted Location of H At Time 2.5

Approx.	x(2.5)	y(2.5)
0	3.38098	2.47759
1	1.93764	2.72562
2	1.48932	1.53066
3	1.37202	1.21823
4	1.34124	1.12519
5	1.33503	1.10598
True	1.33494	1.10571

Repeating the experiment with more closely spaced observations, we converge to the true solution. The determination of the optimal set of observations is itself an interesting question.

REFERENCES

1. Agranovich, Z. S., and V. A. Marchenko, The Inverse Problem of Scattering Theory, Gordon and Breach, New York, 1963.

2. Cannon, J. R., "Determination of Certain Parameters in Heat Conduction Problems," J. Math. Anal. Appl., Vol. 8, No. 2, pp. 188-201, 1964.

3. Maslennikov, M. V., "Uniqueness of the Solution of the Inverse Problem of the Asymptotic Theory of Radiation Transfer," J. Numerical Analysis and Mathematical Physics, Vol. 2, pp. 1044-1053, 1962.

4. Preisendorfer, R. W., "A Survey of Hydrologic Optics," J. Quant. Spectros. Radiat. Transfer, Vol. 8, pp. 325-338, 1968.

5. Sims, A. R., "Certain Aspects of the Inverse Scattering Problem," J. Soc. Indus. Appl. Math., Vol. 5, pp. 183-205, 1957.

6. Kalaba, R. E., and K. Spingarn, "Optimal Inputs and Sensitivities for Parameter Estimation," J. Optimiz. Th. Appl., Vol. 11, pp. 56-67, 1973.

7. Page, William A., Ralph E. Sutton, and Robert J. Miller, "Band Absorption in Nonhomogeneous Isotropically Scattering Planetary Atmospheres - Theory and Experiment," to appear in J. Quant. Spectros. Radiat. Transfer, 1973.

8. Sage, Andrew P., and James L. Melsa, System Identification, Academic Press, New York, 1971.

9. Eykhoff, P., System Parameter and State Estimation, Wiley New York, 1971.

10. Lee, R.C.K., and Y. C. Ho, Optimal Estimation, Identification and Control, M.I.T. Press, Cambridge, Mass., 1964.

11. Bellman, R., H. Kagiwada and R. Kalaba, "Quasilinearization, System Identification, and Prediction," Int. J. Eng. Sci., Vol. 3, pp. 327-334, 1965.

12. Detchmendy, D. M., and R. Sridhar, "Sequential Estimation of States and Parameters in Noisy Nonlinear Dynamical Systems," Trans. ASME, J. Basic Engin., Vol. D88, pp. 362-368, 1966.

13. Kagiwada, H., R. Kalaba, and B. Vereeke, "Invariant Imbedding and Fredholm Integral Equations with Displacement Kernels on an Infinite Interval," Int. J. Comput. Math., Vol. 2, pp. 221-229, 1970.

14. Huss, R., H. Kagiwada, and R. Kalaba, "A Cauchy System for the Green's Function and the Solution of a Two-Point Boundary Value Problem," J. Franklin Institute, Vol. 291, pp. 159-168, 1971.

15. Kagiwada, H., R. Kalaba, and C. Yang, "Reduction of a Class of Nonlinear Integral Equations to a Cauchy System," J. Math. Phys., Vol. 13, pp. 228-231, 1972.

16. Buell, J., H. Kagiwada, R. Kalaba, E. Ruspini, and E. Zagustin, "Solution of a System of Dual Integral Equations," Int. J. Eng. Sci., Vol. 10, pp. 503-510, 1972.

17. Kagiwada, H., and R. Kalaba, "An Initial Value Method for Fredholm Resolvents of Semidegenerate Kernels," J. Optimiz. Th. Appl., Vol. 11, No. 5, pp. 517-532, 1973.

18. Casti, J., and R. Kalaba, Imbedding Methods in Applied Mathematics, Addison-Wesley Publishing Co., Reading, Mass., 1973.

19. Bellman, R., R. Kalaba, and G. M. Wing, "Invariant Imbedding and Mathematical Physics - I: Particle Processes," J. Math. Phys., Vol. 1, pp. 280-308, 1960.

20. Bellman, R., R. Kalaba, and M. C. Prestrud, Invariant Imbedding and Radiative Transfer in Slabs of Finite Thickness, American Elsevier Publishing Co., New York, 1963.

21. Bellman, R., and R. Kalaba, Quasilinearization and Boundary Value Problems, American Elsevier Publishing Co., New York, 1965.

22. Kalaba, R., "On Nonlinear Differential Equations, the Maximum Operation, and Monotone Convergence," J. Math. and Mech., Vol. 8, pp. 519-574, 1959.

23. Bellman, R., Dynamic Programming, Princeton University Press, Princeton, New Jersey, 1957.

24. Bellman, R., H. Kagiwada, and Robert Kalaba, "Orbit Determination as a Multi-Point Boundary-Value Problem and Quasilinearization," Proc. Nat. Acad. Sci., Vol. 48, pp. 1327-1329, 1962.

25. Bellman, R., "Successive Approximations and Computer Storage Problems in Ordinary Differential Equations," Comm. of the ACM, Vol. 4, pp. 222-223, 1961.

26. Bellman, R., H. Kagiwada, R. Kalaba, and M. C. Prestrud, Invariant Imbedding and Time-Dependent Transport Processes, American Elsevier Publishing Co., Inc., New York, 1964.

27. Bellman, R., H. Kagiwada, and R. Kalaba, "Wengert's Method for Partial Derivatives, Orbit Determination and Quasilinearization," Comm. of the ACM, Vol. 8, pp. 231-232, 1965.

28. Bellman, R., H. Kagiwada, R. Kalaba, and R. Sridhar, "Invariant Imbedding and Nonlinear Filtering Theory," J. of the Astronautical Sciences, Vol. 13, No. 3, pp. 110-115, 1966.

29. Henrici, P., Discrete Variable Methods in Ordinary Differential Equations, John Wiley and Sons, New York, 1962.

30. Milne, W. E., Numerical Solution of Differential Equations, John Wiley and Sons, New York, 1953.

31. Lyttleton, R. A., "A Short Method for the Discovery of Neptune," Monthly Notices of the Royal Astronomical Society of London, Vol. 118, pp. 551-559, 1958.

32. Dubyago, A. D., The Determination of Orbits, The Macmillan Company, New York, 1961 (Translation by R. D. Burke, G. Gordon, L. N. Rowell, and F. T. Smith).

CHAPTER 2

IDENTIFICATION OF LAYERED MEDIA WITH MULTIPLE SCATTERING DATA

2-1 INTRODUCTION

Some inverse problems in radiative transfer are concerned with the estimation of the optical properties of an atmosphere based on measurements of diffusely reflected radiation. The location and the intensity of the source of radiation are known. We consider a plane-parallel medium which is composed of two layers. Our aim is to determine the optical thickness and the albedo of each layer, from knowledge of the input radiation and the diffusely reflected light.

First we discuss the concept of invariant imbedding, and we apply this technique to the derivation of the equation for the diffuse reflection function of an inhomogeneous slab with isotropic scattering. The inverse problem is stated in terms of the reflection function, and we formulate the problem as a nonlinear boundary value problem. We then show how the formalism of quasilinearization can be used to solve this problem. We conduct several numerical experiments for the determination of optical thicknesses and albedos of the layers. Computational results are presented, and the FORTRAN computer programs which produced the results are given in Appendix B.

2-2 INVARIANT IMBEDDING

The traditional approach to wave and particle transport processes leads to linear functional equations with boundary conditions. While linearity enables eigenfunction expansions to be made, one finds great difficulty in analytically solving the equations of transfer for all but the simplest cases. Also those equations may have numerically unstable solutions.

The integration of ordinary differential equations with given initial conditions is done extremely efficiently by digital computers, if unstable solutions can be avoided. This suggests that problems be formulated in just this way, with the physical situation as the guide. Invariant imbedding provides a flexible manner in which to relate properties of one process to those of neighboring processes. This also leads to the generalized semigroup concept [1].

In a particle process, one is led by invariant imbedding to differential-integral equations for reflection and transmission functions. By the use of quadrature formulas [2], one reduces the equations from integral-differential form to approximate systems of ordinary differential equations. The wave equation, on the other hand, may be reduced to a system of ordinary differential equations in at least two ways: (1) assume the time factor of the form $e^{i\omega t}$ and the problem simplifies in the steady state situation, or (2) use Laplace transform methods. Both alternatives are discussed in later sections, and other methods of reduction are possible.

Invariant imbedding is a useful formalism, theoretically and computationally speaking. Principles of invariance were first introduced by Ambarzumian in 1943 [3] and developed by Sobolev [4] and Chandrasekhar [5]. The invariance concept was extended and generalized by Bellman and Kalaba [6-8]. Further advances have been made by Kalaba and Kagiwada [9, 10] and others [11]. The form in which invariant imbedding is applied in these chapters is indicated by this example. Suppose that a neutron multiplication process takes place in a rod of length x [12]. The investigator wishes to know the reflected flux r for an input of one particle per second. Rather than study the processes within the rod extending from 0 to x, which would be quite difficult, the experimenter would like to vary the length of the rod and see how the reflected flux changes. The rod length is made a variable of the problem, so that $r = r(x)$. The original situation is imbedded in a class of similar cases, for all lengths of rod, and one obtains directly the reflected flux for a rod of any length including the length under investigation. This flux is rather easily computed and it is physically meaningful [13, 14].

2-3 THE DIFFUSE REFLECTION FUNCTION FOR AN INHOMOGENEOUS SLAB

Consider an inhomogeneous, plane-parallel, non-emitting and isotropically scattering atmosphere of finite optical thickness τ_1. The optical properties depend only on τ, the optical distance from the lower boundary $(0 \leq \tau \leq \tau_1)$. The physical situation is sketched in Fig. 2-1. Parallel rays of light of net flux π per unit area normal to their direction of propagation are incident on the upper surface, $\tau = \tau_1$. The direction is characterized by the parameter μ_0 $(0 < \mu_0 \leq 1)$, which is the cosine of the angle measured from the downward normal to the surface. The bottom surface is a completely absorbing boundary, so that no light is reflected from it. This assumption is not essential to our discussion.

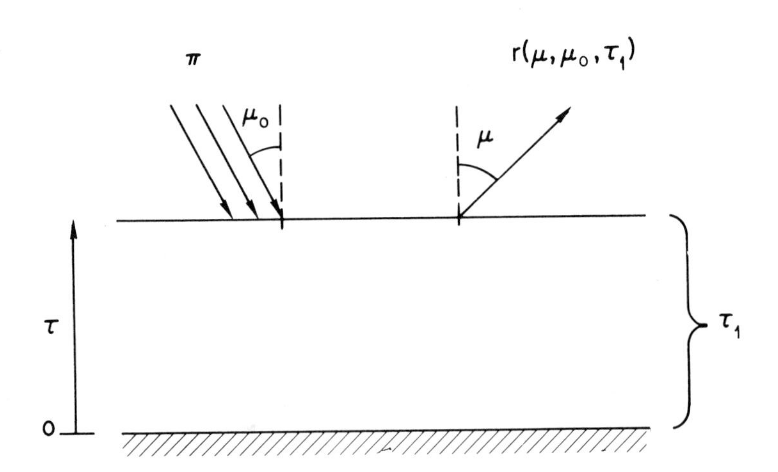

Figure 2-1 Incident and Reflected Rays for an Inhomogeneous Slab of Optical Thickness τ_1 .

The direction of the outgoing radiation is characterized by μ, the cosine of the polar angle measured from the outward normal to the top surface. This parameter is the direction cosine of the vector representing the ray of light. The azimuth angle has no significance due to the symmetry of the situation.

By "intensity" we shall mean the amount of energy which is transmitted through an element of area $d\sigma$ normal to the direction of flow, in a truncated elementary cone $d\omega$ in time dt. See Fig. 2-2, as well as Kourganoff [15]. We restrict ourselves to the steady-state situation at a fixed frequency.

We define the diffuse reflection function as follows:

> $r(\mu,\mu_0,\tau_1)$ is the intensity of the diffusely reflected light in the direction whose cosine is μ with respect to the outward normal, due to incident uniform parallel rays of radiation of constant net flux π in the direction whose cosine is μ_0 with respect to the inward normal, the slab having optical thickness τ_1. (2.1)

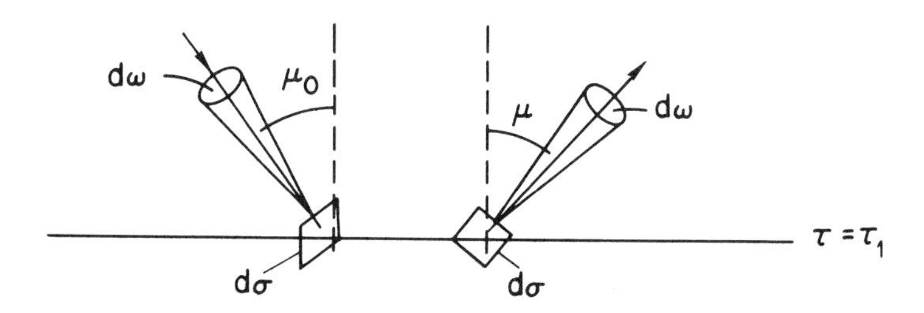

Figure 2-2 The Incident and Reflected Intensities

We define a related function ρ

$$\rho(\mu,\mu_0,\tau_1) = \frac{\mu\, r(\mu,\mu_0,\tau_1)}{\mu_0\, \pi} , \tag{2.2}$$

which is the energy of the diffusely reflected light in the direction μ passing through a unit of horizontal area per unit solid angle per unit time, due to incident radiation of unit energy per unit horizontal area per unit solid angle per unit time, in the direction μ_0. We may also interpret ρ as the probability that a particle will emerge from a unit of horizontal area at $\tau = \tau_1$, the top of a slab of thickness τ_1, going in direction μ, per unit solid angle per unit time, due to an input of one particle per unit horizontal area per unit solid angle per unit time in the direction μ_0.

Consider now a slab of thickness $\tau_1 + \Delta$ formed by adding a thin slab of thickness Δ to the top of the slab of thickness τ_1, as illustrated in Fig. 2-3. By imbedding the original problem in a class of problems of similar nature, we will derive an integro-differential equation for the diffuse reflection function.

The diffuse reflection function for a slab of thickness $\tau_1 + \Delta$ with an input of net flux π is $r(\mu,\mu_0,\tau_1{+}\Delta) = \pi\rho(\mu,\mu_0,\tau_1{+}\Delta)\mu_0/\mu$. Applying the method of invariant imbedding in its particle counting form to the probability of emergence of a particle from a slab, we obtain the equation

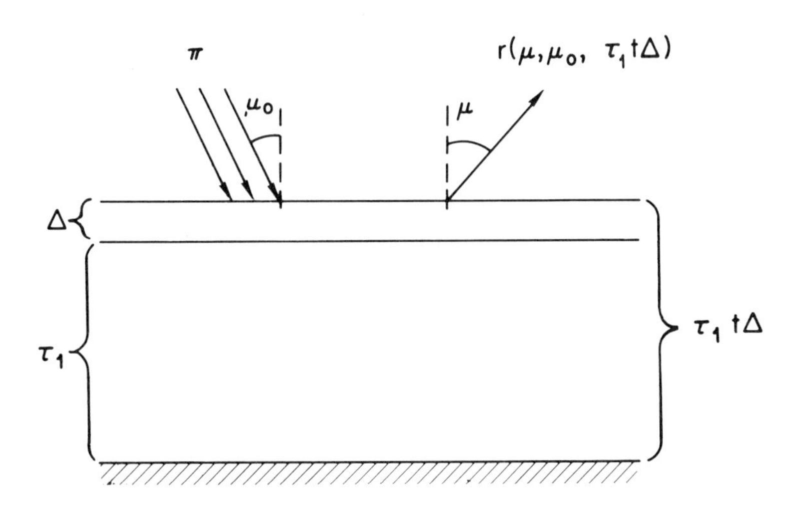

Figure 2-3 An Inhomogeneous Slab of Optical Thickness $\tau_1 + \Delta$

$$
\begin{aligned}
\rho(\mu,\mu_0,\tau_1+\Delta) &= \rho(\mu,\mu_0,\tau_1) - \Delta \left(\frac{1}{\mu_0} + \frac{1}{\mu}\right) \rho(\mu,\mu_0,\tau_1) \\
&+ \frac{\Delta}{\mu_0} \frac{\lambda(\tau_1)}{4\pi} + 2\pi \int_0^1 \rho(\mu',\mu_0,\tau_1)d\mu' \frac{\Delta}{\mu'} \frac{\lambda(\tau_1)}{4\pi} \\
&+ 2\pi \int_0^1 \frac{\Delta}{\mu_0} \frac{\lambda(\tau_1)}{4\pi} d\mu'' \rho(\mu,\mu'',\tau_1) \qquad (2.3) \\
&+ 2\pi \int_0^1 \rho(\mu',\mu_0,\tau_1)d\mu' \frac{\Delta}{\mu'} \frac{\lambda(\tau_1)}{4\pi} \\
&\cdot 2\pi \int_0^1 \rho(\mu,\mu'',\tau_1)d\mu'' + o(\Delta).
\end{aligned}
$$

The first term on the right-hand side is the probability that a particle emerges without any interaction in the thin slab. The unit of distance is such that x is the probability of an interaction in a path of length x. Hence the second term represents the losses due to interactions of the incoming and

and outgoing particles whose path lengths in Δ are Δ/μ_0 and Δ/μ respectively. The third term is the probability of an interaction and re-emission isotropically into the direction given by μ. The function $\lambda(\tau_1)$ is the probability of re-emission, and is called the albedo for single scattering. The next term is the probability that the particle is diffusely reflected from the slab between $(0,\tau_1)$ into the direction μ' and interacts in Δ and is re-emitted into the direction of emergence μ. The next term is the probability that an incoming particle interacts in Δ, enters the slab $(0,\tau_1)$ and is diffusely scattered into the emergent direction μ. The sixth term is the probability of diffuse reflection in $(0,\tau_1)$, then interaction and re-emission in Δ, and diffuse reflection from $(0,\tau_1)$ with outgoing direction μ. All other probabilities are proportional to Δ^2 or higher powers of Δ and are accounted for in the term $o(\Delta)$.

Let the diffusely reflected intensity be given by a new function R, by means of the relation

$$r(\mu,\mu_0,\tau_1) = \frac{R(\mu,\mu_0,\tau_1)}{4\pi}, \tag{2.4}$$

where R is related to ρ by the formula

$$\rho(\mu,\mu_0,\tau_1) = \frac{R(\mu,\mu_0,\tau_1)}{4\pi\mu_0}. \tag{2.5}$$

Then R satisfies the equation

$$
\begin{aligned}
R(\mu,\mu_0,\tau_1+\Delta) &= R(\mu,\mu_0,\tau_1) - \Delta\left(\frac{1}{\mu_0}+\frac{1}{\mu}\right) R(\mu,\mu_0,\tau_1) \\
&+ \Delta\,\lambda\left\{1 + \frac{1}{2}\int_0^1 R(\mu',\mu_0,\tau_1)\,\frac{d\mu'}{\mu'}\right. \\
&+ \frac{1}{2}\int_0^1 R(\mu,\mu'',\tau_1)\,\frac{d\mu''}{\mu''} \qquad (2.6) \\
&+ \left.\frac{1}{4}\int_0^1 R(\mu',\mu_0,\tau_1)\frac{d\mu'}{\mu'}\int_0^1 R(\mu,\mu'',\tau_1)\frac{d\mu''}{\mu''}\right\} + o(\Delta).
\end{aligned}
$$

We expand the left-hand side of the equation in powers of Δ,

$$
R(\mu,\mu_0,\tau_1+\Delta) = R(\mu,\mu_0,\tau_1) + \frac{\partial R(\mu,\mu_0,\tau_1)}{\partial\tau_1}\,\Delta + o(\Delta). \qquad (2.7)
$$

By letting $\Delta\to 0$, we arrive at the integro-differential equation

$$
\begin{aligned}
&\frac{\partial R(\mu,\mu_0,\tau_1)}{\partial\tau_1} + \left(\frac{1}{\mu_0}+\frac{1}{\mu}\right) R(\mu,\mu_0,\tau_1) \\
&\quad = \lambda(\tau_1)\left[1 + \frac{1}{2}\int_0^1 R(\mu',\mu_0,\tau_1)\frac{d\mu'}{\mu'}\right] \qquad (2.8) \\
&\quad \cdot\left[1 + \frac{1}{2}\int_0^1 R(\mu,\mu'',\tau_1)\frac{d\mu''}{\mu''}\right].
\end{aligned}
$$

The initial condition is

$$
R(\mu,\mu_0,0) = 0, \qquad (2.9)
$$

because no light is diffusely reflected when the medium has zero thickness. It is readily seen that the function R is symmetric [5, 16-18], i.e.,

$$R(\mu,\mu_0,\tau_1) = R(\mu_0,\mu,\tau_1). \tag{2.10}$$

Equation (2.8) for R is the same as Chandrasekhar's equation for his scattering function S, when the medium is homogeneous and isotropic [5]. It is a Ricatti equation.

2-4 GAUSSIAN QUADRATURE

The above integrals may be evaluated by the use of Gaussian quadrature [5, 16, 17]. Since the limits of our integrals are zero to one, we use the approximate relation

$$\int_0^1 f(x)\,dx \cong \sum_{k=1}^{N} f(a_k)w_k, \tag{2.11}$$

which is exact if $f(x)$ is a polynomial of degree $2N-1$ or less. The numbers a_k are roots of the shifted Legendre function $P_N^*(x) = P_N(1-2x)$ on the interval $(0,1)$, and the numbers w_k are the corresponding weights. For a more detailed discussion and for tables of roots and weights, see [16].

Replacing integrals by Gaussian sums, we have the following equation,

$$\frac{\partial R(\mu,\mu_0,\tau_1)}{\partial \tau_1} + \left(\frac{1}{\mu_0} + \frac{1}{\mu}\right) R(\mu,\mu_0,\tau_1)$$

$$= \lambda(\tau_1) \left[1 + \frac{1}{2} \sum_{k=1}^{N} R(\mu_k,\mu_0,\tau_1) \frac{w_k}{\mu_k}\right] \tag{2.12}$$

$$\cdot \left[1 + \frac{1}{2} \sum_{k=1}^{N} R(\mu,\mu_k,\tau_1) \frac{w_k}{\mu_k}\right].$$

For $N \sim 7$, this is a fairly good approximation [16, 17]. We consider only those incident and outgoing directions for which the cosines take on the values of the roots μ_k.

For $N = 7$, the roots μ_k and the corresponding angles, arc cosine μ_k, are listed in Table 2-1 in order of increasing μ.

We define the functions of one argument,

$$R_{ij}(\tau_1) = R(\mu_i,\mu_j,\tau_1), \tag{2.13}$$

for $i = 1, 2, \ldots, N$, $j = 1, 2, \ldots, N$. Then (2.12) becomes a system of ordinary differential equations

Table 2-1

Roots of Shifted Legendre Polynomials of Degree N = 7, and Corresponding Angles

k	Roots μ_k	Arc cosine μ_k (in degrees)
1	0.025446044	88.541891
2	0.12923441	82.574646
3	0.29707742	72.717849
4	0.50000000	60.000000
5	0.70292258	45.338044
6	0.87076559	29.452271
7	0.97455396	12.953079

$$\frac{dR_{ij}(\tau_1)}{d\tau_1} + \left(\frac{1}{\mu_i} + \frac{1}{\mu_j}\right) R_{ij} = \lambda(\tau_1) \cdot \left[1 + \frac{1}{2}\sum_{k=1}^{N} R_{kj}(\tau_1)\frac{w_k}{\mu_k}\right]\left[1 + \frac{1}{2}\sum_{k=1}^{N} R_{ik}(\tau_1)\frac{w_k}{\mu_k}\right], \tag{2.14}$$

with optical thickness τ_1 as the independent variable. The initial conditions are, of course,

$$R_{ij}(0) = 0. \tag{2.15}$$

The system of N^2 first order differential equations reduce to a system of $N(N+1)/2$ equations by the use of the symmetry property of R. This is a large saving of computational effort.

2-5 AN INVERSE PROBLEM

Consider the inhomogeneous medium composed of two layers as illustrated in Fig. 2-4.

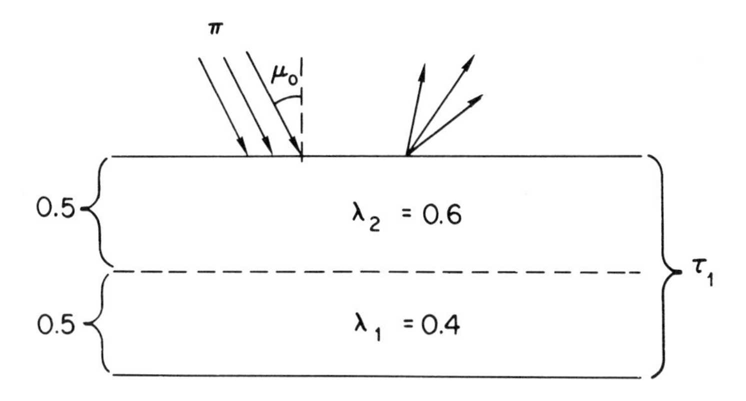

Figure 2-4 A Layered Medium

The total thickness of the medium is 1.0, the thickness of each slab is 0.5, and the albedos are 0.4 in the lower layer, 0.6 in the upper layer. In order to have a continuous function for the albedo, we assume that λ is given by the function

$$\lambda(\tau) = 0.5 + 0.1 \tanh 10(\tau - 0.5). \qquad (2.16)$$

This function is plotted in Fig. 2-5.

Parallel rays of net flux π are incident on the upper surface of the medium in a direction characterized by

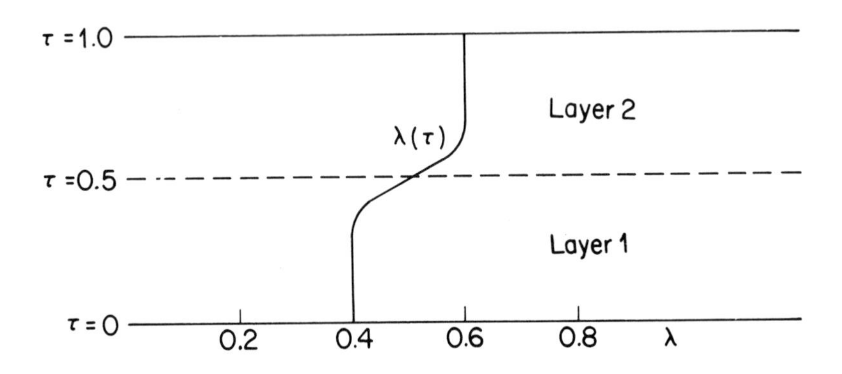

Figure 2-5 The Albedo Function $\lambda(\tau) = 0.5 + 0.1 \tanh 10(\tau-0.5)$ for a Slab of Thickness 1.0.

the parameter μ_j. We obtain N^2 measurements of the intensity of the diffusely reflected light, $b_{ij} \cong r_{ij}(\tau_1)$, for incident directions μ_j, $j = 1, 2, \ldots, N$, and reflection directions μ_i, $i = 1, 2, \ldots, N$. We wish to determine the nature of the medium from the knowledge of the reflected radiation.

Let the total optical thickness of the slab be T, and let the thickness of the lower layer be c. Let the albedos be λ_1 and λ_2, for the lower and upper slabs respectively, where the albedo as a function of optical elevation is

$$\lambda(\tau) = a + b \tanh 10(\tau-c)$$

and

$$\lambda_1 \cong a - b \tag{2.17}$$

$$\lambda_2 \cong a + b,$$

where a and b are unknown parameters. For the "true" situation,

$$T = 1.0, \ a = 0.5, \ b = 0.1, \ c = 0.5.$$

The inverse problem which we wish to solve is to determine the quantities T, a, b, and c in such a way as to have best agreement, in the least square sense, between the estimated solution using the ordinary differential equations for r_{ij} and the observed reflection pattern. Mathematically speaking, we wish to minimize the expression

$$\sum_{i=1}^{N} \sum_{j=1}^{N} \left[r_{ij}(T) - b_{ij}\right]^2 \tag{2.18}$$

over all choices of the unknown parameters.

In Table 2-2, we present the measurements $\{b_{ij}\}$ for $N = 7$. In Fig. 2-6 we plot some of the measurements as a function of the cosine of the reflection angle, $\mu \sim \mu_i$, for input directions $\mu_0 \sim \mu_j \cong .025$, $.5$, and $.975$. The discrete observations are shown as dots, and for clarity we draw smooth curves through these points. For comparison, we show what the corresponding measurements would be if the medium were homogeneous with albedo $\lambda = 0.5$. There are distinct differences between corresponding curves.

Table 2-2

The Measurements $\{b_{ij}\}$

	i = 1	2	3	4	5	6	7
j = 1	0.079914	0.028164	0.014304	0.009104	0.006707	0.005515	0.004970
2	0.143038	0.091522	0.058437	0.040826	0.031405	0.026378	0.023989
3	0.167000	0.134331	0.099653	0.075106	0.060044	0.051445	0.047248
4	0.178898	0.157955	0.126408	0.099392	0.081253	0.070435	0.065042
5	0.185284	0.170817	0.142072	0.114229	0.094495	0.082423	0.076332
6	0.188723	0.177733	0.150791	0.122665	0.102104	0.089349	0.082870
7	0.190354	0.180898	0.154995	0.126773	0.105829	0.092748	0.086083

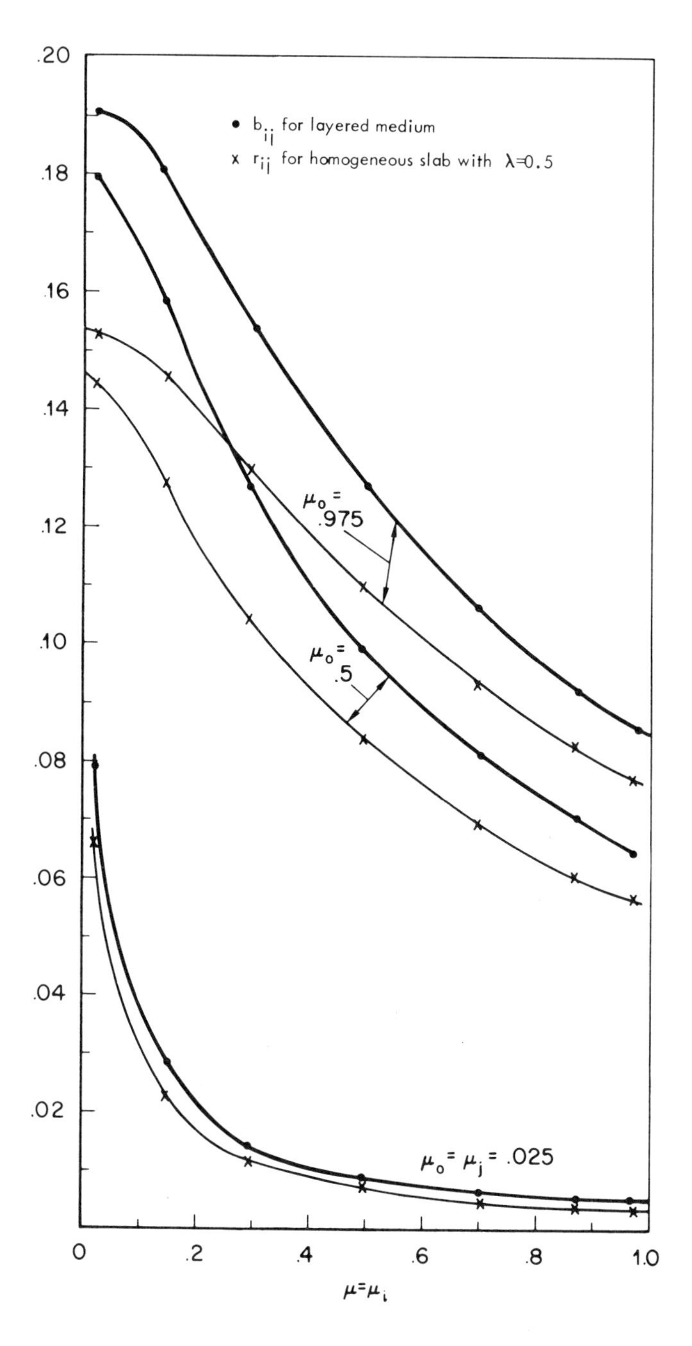

Figure 2-6 Some of the Measurements (b_{ij}) for a Layered Medium

2-6 FORMULATION AS A NONLINEAR BOUNDARY VALUE PROBLEM

We formulate this inverse problem as a nonlinear boundary value problem. To the system of N^2 nonlinear differential equations

$$\frac{dR_{ij}}{d\tau_1} + \left(\frac{1}{\mu_i} + \frac{1}{\mu_j}\right) R_{ij} = \lambda(\tau_1) \cdot \left[1 + \frac{1}{2}\sum_{k=1}^{N} R_{kj}\frac{w_k}{\mu_k}\right]\left[1 + \frac{1}{2}\sum_{k=1}^{N} R_{ik}\frac{w_k}{\mu_k}\right], \tag{2.19}$$

where

$$\lambda(\tau_1) = a + b \tanh 10(\tau_1 - c), \tag{2.20}$$

we add the differential equations

$$\frac{da}{d\tau_1} = 0, \quad \frac{db}{d\tau_1} = 0, \quad \frac{dc}{d\tau_1} = 0, \quad \frac{dT}{d\tau_1} = 0, \tag{2.21}$$

because a, b, and c and T are unknown constants. The boundary conditions are

$$R_{ij}(0) = 0, \tag{2.22}$$

and

$$\frac{\partial S}{\partial a} = 0, \quad \frac{\partial S}{\partial b} = 0, \quad \frac{\partial S}{\partial c} = 0, \quad \frac{\partial S}{\partial T} = 0, \tag{2.23}$$

where

$$S = \sum_{i=1}^{N} \sum_{j=1}^{N} [R_{ij}(T) - 4\mu_i b_{ij}]^2. \tag{2.24}$$

2-7 NUMERICAL EXPERIMENTS I. DETERMINATION OF c, THE THICKNESS OF THE LOWER LAYER

Let us try to determine the quantity c, which is the thickness of the lower layer of the stratified medium. We assume that all of the other parameters a, b, and T are known. The parameter c is considered to be a function of optical height τ_1 described by the equation $dc/d\tau_1 = 0$. By following the method of quasilinearization as described previously, we obtain a system of linear differential equations for the $(k+1)^{st}$ approximation to R_{ij} and c:

$$\frac{dR_{ij}^{k+1}}{d\tau_1} = f(R_{ij}^k, c^k) + \sum_{i,j} \left[R_{ij}^{k+1} - R_{ij}^k\right] \frac{\partial f}{\partial R_{ij}^k}$$

$$+ (c^{k+1} - c^k) \frac{\partial f}{\partial c^k}, \tag{2.25}$$

$$\frac{dc^{k+1}}{d\tau_1} = 0 .$$

where

$$f(R_{ij}^k, c^k) = -\left(\frac{1}{\mu_i} + \frac{1}{\mu_j}\right) R_{ij}^k + \lambda(c^k)\left[1 + \frac{1}{2}\sum_{\ell=1}^{N} R_{\ell j}^k \frac{w_\ell}{\mu_\ell}\right]$$

$$\cdot\left[1 + \frac{1}{2}\sum_{\ell=1}^{N} R_{i\ell}^k \frac{w_\ell}{\mu_\ell}\right], \tag{2.26}$$

$$\lambda(c^k) = a + b \tanh 10(\tau_1 - c^k). \tag{2.27}$$

After simplifying, we have

$$\frac{dR_{ij}^{k+1}}{d\tau_1} = \left\{-\left(\frac{1}{\mu_i} + \frac{1}{\mu_j}\right) R_{ij}^k + \lambda(c^k)\left(1 + \frac{1}{2}\sum_{\ell=1}^{N} R_{i\ell}^k \frac{w_\ell}{\mu_\ell}\right)\right.$$

$$\left.\cdot\left(1 + \frac{1}{2}\sum_{\ell=1}^{N} R_{\ell j}^k \frac{w_\ell}{\mu_\ell}\right)\right\} + \left\{-\left(\frac{1}{\mu_i} + \frac{1}{\mu_j}\right)\left(R_{ij}^{k+1} - R_{ij}^k\right)\right.$$

$$+ \frac{1}{2}\lambda(c^k)\left[\left(1 + \frac{1}{2}\sum_{\ell=1}^{N} R_{i\ell}^k \frac{w_\ell}{\mu_\ell}\right) \cdot \sum_{\ell=1}^{N}\left(R_{\ell j}^{k+1} - R_{\ell j}^k\right)\frac{w_\ell}{\mu_\ell}\right.$$

$$\left.\left. + \left(1 + \frac{1}{2}\sum_{\ell=1}^{N} R_{\ell j}^k \frac{w_\ell}{\mu_\ell}\right) \cdot \sum_{\ell=1}^{N}\left(R_{i\ell}^{k+1} - R_{i\ell}^k\right)\frac{w_\ell}{\mu_\ell}\right]\right\}$$

$$+ \left\{(c^{k+1} - c^k)\left(1 + \frac{1}{2}\sum_{\ell=1}^{N} R_{i\ell}^k \frac{w_\ell}{\mu_\ell}\right)\left(1 + \frac{1}{2} R_{\ell j}^k \frac{w_\ell}{\mu_\ell}\right)\right.$$

$$\left.\cdot \left(-10\, b\, \mathrm{sech}^2\, 10(\tau_1 - c^k)\right)\right\},$$

$$\frac{dc^{k+1}}{d\tau_1} = 0\ . \tag{2.28}$$

Since $N = 7$, there are basically $7^2 + 1 = 50$ differential equations, which reduce to $7 \cdot 8/2 + 1 = 29$ differential equations by the use of the symmetry property

$$R_{ij}^{k+1}(\tau_1) = R_{ji}^{k+1}(\tau_1). \qquad (2.29)$$

While the computations are reduced, the full set of values R_{ij}^{k+1} representing a 7 x 7 matrix is always available.

Now let the 50-dimensional vector $x^{k+1}(\tau_1)$ have the components

$$x_{\ell}^{k+1}(\tau_1) = R_{ij}^{k+1}(\tau_1), \qquad (2.30)$$

for $\ell = 1, 2, \ldots, 49$ as $i = 1, 2, \ldots, 7$ and $j = 1, 2, \ldots, 7$, and

$$x_{50}^{k+1}(\tau_1) = c^{k+1}(\tau_1). \qquad (2.31)$$

Since $x^{k+1}(\tau_1)$ is a solution of a system of linear differential equations, we may represent it as the sum of a particular vector solution, $p(\tau_1)$, and a vector solution of the homogeneous system, $h(\tau_1)$,

$$x^{k+1}(\tau_1) = p(\tau_1) + m\, h(\tau_1). \qquad (2.32)$$

The system of differential equations for $p(\tau_1)$ is obtained by substituting the appropriate component of p wherever R^{k+1} or c^{k+1} occurs in (2.28). We choose the initial conditions $p(0) = 0$. The system of equations for the homogeneous solution is similarly obtained, but of course all terms not involving the $(k+1)^{st}$ approximation are dropped. The initial vector $h(0)$ has all of its components zero except for the last, which is unity. The boundary conditions $R_{ij}^{k+1}(0) = 0$ are identically satisfied. The solutions $p(\tau_1)$ and $h(\tau_1)$ are produced on the interval $0 \leq \tau_1 \leq 1.0$ by numerical integration.

The multiplier m is chosen to minimize the quadratic form,

$$S = \sum_{\ell=1}^{49} \left[p_\ell(1) + m\, h_\ell(1) - b_\ell \right]^2, \tag{2.33}$$

where the observations are $b_\ell \cong x_\ell^{k+1}(1)$. It is required that

$$\frac{\partial S}{\partial m} = 0 \tag{2.34}$$

and so the value of m is

$$m = \frac{\sum_{\ell=1}^{49} h_\ell(1)\,[b_\ell - p_\ell(1)]}{\sum_{\ell=1}^{49} [h_\ell(1)]^2} . \tag{2.35}$$

The thickness of the lower layer in the new approximation is

$$c^{k+1} = m. \tag{2.36}$$

The initial approximation required for this successive approximation scheme is produced by numerically integrating the nonlinear system of equations for R using a rough estimate of c. The results of three experiments with initial guesses c = 0.2, 0.8, and 0.0, respectively, are given in Table 2-3. The values of c obtained in the first, second, third and fourth approximations are tabulated.

Table 2-3

Successive Approximations of c,
The Level of the Interface

Approximation	Run 1	Run 2	Run 3
0	0.2	0.8	0.0
1	0.62	0.57	
2	0.5187	0.5024	No convergence
3	0.500089	0.499970	
4	0.499990	0.499991	
True Value	0.5	0.5	0.5

The initial guess of c in Run 1 is 60% too low, and in Run 2, 60% too high. Yet the correct value of c is accurately found in 3 to 4 iterations. The time required for each run is about 2 minutes on the IBM 7044 digital computer, using an Adams-Moulton fourth order integration scheme with a grid size of $\Delta\tau_1 = 0.01$. Each iteration requires the integration of $2 \times 29 = 58$ differential equations with initial values, and the values of $p_\ell(\tau_1)$ and $h_\ell(\tau_1)$ thus produced are stored in the rapid access memory of the computer at each of a hundred and one grid points, $\tau_1 = 0, .01, .02,$

..., 1.0. The current approximation of R_{ij}^k is also stored at a hundred and one points.

Run 3 is an unsuccessful experiment because the initial guess for c, i.e., a single layer approximation, is very poor. The solution diverges.

2-8 NUMERICAL EXPERIMENTS II. DETERMINATION OF T, THE OVERALL OPTICAL THICKNESS

Now let us try to estimate the total optical thickness T of the stratified medium, assuming that we know all of the other parameters of the system. Again we are provided with 49 measurements of $\{b\}$, the intensity of the diffusely reflected radiation in various directions.

The quantity T is the end point of the range of integration, i.e., $0 \leqq \tau_1 \leqq T$. In order to have a <u>known</u> end point, we define a new independent variable σ,

$$\sigma T = \tau_1 \ , \tag{2.37}$$

so that the integration interval is fixed, $0 \leq \sigma \leq 1$. Then T satisfies the equation, $dT/d\sigma = 0$. Our system of nonlinear equations is

$$\frac{dR_{ij}(\sigma)}{d\sigma} = T \left\{ -\left(\frac{1}{\mu_i} + \frac{1}{\mu_j}\right) R_{ij} + \lambda \left[1 + \frac{1}{2} \sum_{k=1}^{N} R_{ik} \frac{w_k}{\mu_k}\right] \left[1 + \frac{1}{2} \sum_{k=1}^{N} R_{kj} \frac{w_k}{\mu_k}\right] \right\} , \tag{2.38}$$

$$\frac{dT}{d\sigma} = 0 \ .$$

where $\lambda = a + b \tanh 10(\sigma T - c)$.

The solution is subject to the conditions

$$R_{ij}(0) = 0, \tag{2.39}$$

$$\min_{T} \sum_{i=1}^{N} \sum_{j=1}^{N} [R_{ij}(T) - 4\mu_i b_{ij}]^2 . \tag{2.40}$$

Linear differential equations are obtained in the same manner as before, and we solve a sequence of linear boundary value problems.

Three trials are made to determine the thickness T, with initial guesses $T = 0.9$, 1.5, and 0.5, while the correct value is 1.0. Four iterations yield a value of T which is correct to one part in a hundred thousand, in each of the three experiments. The total computing time is four minutes. The experiment is successful even when the initial guess is only one-half of the true value.

2-9 NUMERICAL EXPERIMENTS III. DETERMINATION OF THE TWO ALBEDOS AND THE THICKNESS OF THE LOWER LAYER

Given 49 measurements of the diffusely reflected light, we wish to determine the two albedos

$$\lambda_1 \cong a - b, \quad \lambda_2 \cong a + b , \tag{2.41}$$

and the thicknesses of the two layers. We assume that we know the overall thickness $T = 1.0$, and so if the thickness of the lower layer is c, the thickness of the upper layer is given by T - c. The unknown parameters are a, b, and c. Since there are three unknowns, we have three homogeneous solutions and of course a particular solution to compute in each iteration of the experiment. Each solution has 28 + 3 = 31 components, so that there are 4 x 31 = 124 linear differential equations being integrated during each stage of the quasilinearization scheme. The three multipliers form the solution of a third order linear algebraic system. They are found by a matrix inversion using a Gaussian elimination method. Table 2-4 summarizes the results of an experiment which is carried out in about 2 minutes on the IBM 7044. The FORTRAN IV computer programs for all three series of experiments are given in Appendix B.

Table 2-4

Successive Approximations of λ_1, λ_2, and c

Approximation	$\lambda_1 = a-b$	$\lambda_2 = a+b$	c
0	0.51	0.69	0.4
1	0.4200	0.6052	0.5038
2	0.399929	0.599995	0.499602
3	0.399938	0.599994	0.499878
True Value	0.4	0.6	0.5

2-10 DISCUSSION

The approach which is discussed above is readily extended to other inverse problems with different physical situations. The numerical experiments in this chapter make use of many accurate observations of the reflected light while in the next chapter, the effect of errors in the measurements is examined. We note that initial approximations must be good enough to insure convergence. A rational initial estimate may be made from knowledge of the diffuse reflection fields for various homogeneous slabs, as calculated for example by Bellman, Kalaba and Prestrud [17]. Other inverse problems might deal with the transmission function, the source function, the X and Y functions, and the emergence probabilities [19-22].

REFERENCES

1. Hille, E., and R. Phillips, Functional Analysis and Semi-Groups, American Mathematical Society, Providence, R.I., 1957.

2. Frank, Philipp, and Richard Von Mises, Die Differential- und Integralgleichungen, Mary S. Rosenberg, New York, 1943.

3. Ambarzumian, V. A., "Diffuse Reflection of Light by a Foggy Medium," Compt. Rend. Acad. Sci. U.S.S.R., Vol. 38, No. 8, pp. 229-232, 1943.

4. Sobolev, V. V., A Treatise on Radiative Transfer, Cambridge University Press, 1960.

5. Chandrasekhar, S., Radiative Transfer, Oxford University Press, London, 1950.

6. Bellman, R., and R. Kalaba, "On the Principle of Invariant Imbedding and Propagation through Inhomogeneous Media," Proc. Nat. Acad. Sci. USA, Vol. 42, pp. 629-632, 1956.

7. Bellman, R., and R. Kalaba, "Functional Equations, Wave Propagation, and Invariant Imbedding," J. Math. Mech., Vol. 8, pp. 683-704, 1959.

8. Bellman, R., R. Kalaba, and G. M. Wing, "Invariant Imbedding and Mathematical Physics-I: Particle Processes," J. Math. Phys., pp. 280-308, 1960.

9. Kagiwada, H., and R. Kalaba, "An Initial Value Method Suitable for the Computation of Certain Fredholm Resolvents," J. Math. and Phys. Sciences, Vol. 1, No. 1, pp. 109-122, 1967.

10. Kagiwada, H., and R. Kalaba, "The Basic Functions b and h for Fredholm Integral Equations with Displacement Kernels," J. Optimiz. Theory Appl., Vol. 11, No. 5, pp. 517-532, 1973.

11. Scott, M. R., A Bibliography on Invariant Imbedding and Related Topics, Sandia Laboratories, 1972.

12. Wing, G. M., An Introduction to Transport Theory, John Wiley and Sons, Inc., New York, 1962.

13. Shimizu, A., Application of Invariant Imbedding to Penetration Problems of Neutrons and Gamma Rays, NAIG Nuclear Research Laboratory, Research Memo-1, December 1963.

14. Fujita, H., K. Kobayashi, and T. Hyodo, "Backscattering of Gamma Rays from Iron Slabs," Nuclear Science and Engineering, Vol. 19, pp. 437-440, 1964.

15. Kourganoff, V., Basic Methods in Transfer Problems, Oxford University Press, London, 1952.

16. Bellman, R. E., H. H. Kagiwada, R. E. Kalaba, and M. C. Prestrud, Invariant Imbedding and Time-Dependent Transport Processes, American Elsevier Publishing Co., Inc., New York, 1964.

17. Bellman, R., R. Kalaba, and M. Prestrud, Invariant Imbedding and Radiative Transfer in Slabs of Finite Thickness, American Elsevier Publishing Co., Inc., New York, 1963.

18. Minnaert, M., "The Reciprocity Principle in Lunar Photometry," Astrophys. J., Vol. 93, pp. 403-410, 1941.

19. Bellman, R., H. Kagiwada, R. Kalaba, and S. Ueno, "A New Derivation of the Integro-Differential Equations for Chandrasekhar's X and Y Functions," J. Math. Phys., Vol. 9, pp. 906-908, 1968.

20. Bellman, R., H. Kagiwada, R. Kalaba, and S. Ueno, "Invariant Imbedding and the Computation of Internal Fields for Transport Processes," J. Math. Anal. Appl., Vol. 12, pp. 541-548, 1965.

21. Busbridge, I. W., The Mathematics of Radiative Transfer, Cambridge University Press, 1960.

22. Sobolev, V. V., Rasseyanie sveta v atmosferakh planet, Moscow, 1972.

CHAPTER 3

IDENTIFICATION USING NOISY SCATTERING MEASUREMENTS

3-1 INTRODUCTION

The techniques of invariant imbedding and quasilinearization are applied to some inverse problems of radiative transfer through an inhomogeneous slab in which the albedo for single scattering has a parabolic dependence on optical height. The results of many numerical experiments on the effect of the angle of incidence of radiation, errors in observations, and minimax versus least squares criterion are reported. Other experiments are carried out to design an optical medium according to specified requirements. The knowledge gained through this type of numerical experimentation should prove useful in the planning of laboratory or satellite experiments as well as for the reduction of data and the construction of model atmospheres.

3-2 AN INVERSE PROBLEM

Consider an inhomogeneous, plane-parallel, non-emitting and isotropically scattering atmosphere of finite optical thickness τ_1. Its optical properties depend only on the optical distance τ from the bottom surface. The bottom surface is a completely absorbing boundary, so that no light is reflected from it. See Fig. 3-1 for a sketch of the physical situation. Parallel rays of light of net flux π per unit area normal to their direction of propagation are incident on the upper surface. The direction is specified by μ_0 $(0 < \mu_0 \leq 1)$, the cosine of the angle measured from the normal to the surface.

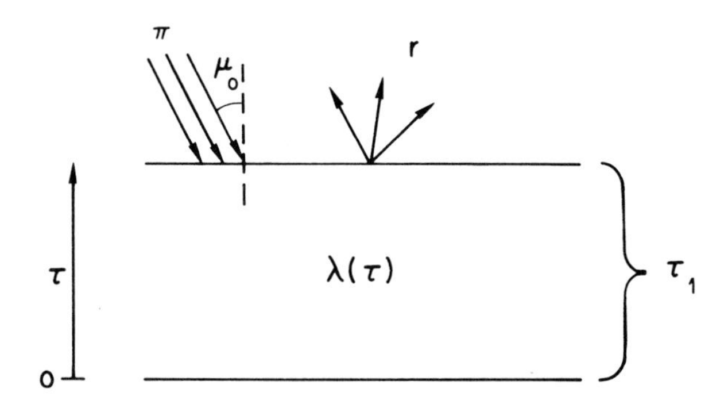

Figure 3-1 The Physical Situation

Let $r(\mu, \mu_0, \tau_1)$ denote the intensity of the diffusely reflected light in the direction μ, and set $R(\mu, \mu_0, \tau_1) = 4\mu r$. Then the function R satisfies the integro-differential equation

$$\frac{\partial R}{\partial \tau_1} = -\left(\frac{1}{\mu} + \frac{1}{\mu_0}\right) R + \lambda(\tau_1) \left[1 + \frac{1}{2}\int_0^1 R(\mu,\mu',\tau_1) \frac{d\mu'}{\mu'}\right]$$

$$\cdot \left[1 + \frac{1}{2}\int_0^1 R(\mu',\mu_0,\tau_1) \frac{d\mu'}{\mu'}\right] \tag{3.1}$$

with initial condition

$$R(\mu,\mu_0,\tau_1) = 0. \tag{3.2}$$

The function $\lambda(\tau_1)$ is the albedo for single scattering.

We wish to consider the inverse problem of estimating the optical properties of the medium as represented by $\lambda(\tau)$ as well as the optical thickness of the slab, based on measurements of the diffusely reflected light.

3-3 FORMULATION AS A NONLINEAR BOUNDARY VALUE PROBLEM

Let us consider the case in which the albedo may be assumed to have a parabolic form,

$$\lambda(\tau) = \frac{1}{2} + a\tau + b\tau^2 , \tag{3.3}$$

where a and b are constants for a particular slab. For example, let us take $a = 2$ and $b = -2$, and we choose the optical thickness $c = 1.0$. The albedo as a function of optical height is shown in Fig. 3-2.

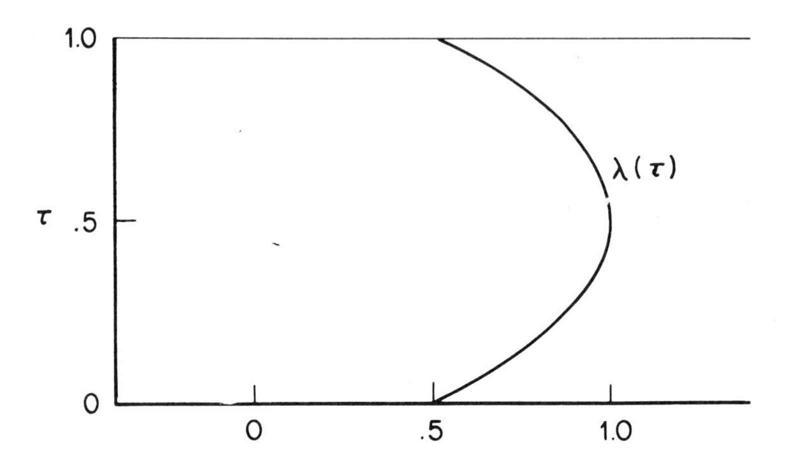

Figure 3-2 A Parabolic Albedo Function, $\lambda(\tau) = \frac{1}{2} + 2\tau - 2\tau^2$, for a Slab of Thickness 1.0

We replace the integro-differential equation by the discrete approximate system obtained by the use of Gaussian quadrature,

$$\frac{dR_{ij}}{d\tau_1} = -\left(\frac{1}{\mu_i} + \frac{1}{\mu_j}\right) + \lambda(\tau_1)\left[1 + \frac{1}{2}\sum_{k=1}^{N} R_{ik}(\tau_1)\frac{w_k}{\mu_k}\right]$$

$$\cdot \left[1 + \frac{1}{2}\sum_{k=1}^{N} R_{kj}(\tau_1)\frac{w_k}{\mu_k}\right]. \tag{3.4}$$

In these equations, $R_{ij}(\tau_1)$ represents $R(\mu_i, \mu_j, \tau_1)$.

We produce "observations" of the diffusely reflected light by choosing $N = 7$, and integrating (3.4) from $\tau_1 = 0$

to $\tau_1 = 1.0$, and then setting $b_{ij} = \frac{R_{ij}}{4\mu_i}$. Then $\{b_{ij}\}$ is the set of measurements for $\tau_1 = 1$.

Starting with the observations $\{b_{ij}\} \cong \{r_{ij}(c)\}$, we wish to determine the quantities a, b, and the optical thickness c which minimize the expression

$$S = \sum_{i,j} [r_{ij}(c) - b_{ij}]^2 , \tag{3.5}$$

where $R_{ij}(\tau_1) = 4\mu_i r_{ij}(\tau_1)$ is the solution of the nonlinear system (3.4). This inverse problem may be viewed as a nonlinear boundary-value problem.

3-4 SOLUTION VIA QUASILINEARIZATION

Since the terminal value of the independent variable τ_1 is unknown, we make the following transformation to a new independent variable σ,

$$\sigma = \tau_1/c , \tag{3.6}$$

which has initial value 0 and terminal value 1.0. Then the parameters a, b, and the thickness c satisfy the equations

$$\frac{da}{d\sigma} = 0, \quad \frac{db}{d\sigma} = 0, \quad \frac{dc}{d\sigma} = 0. \tag{3.7}$$

Eqs. (3.7) are added to the system

$$\frac{dR_{ij}}{d\sigma} = c\left\{-\left(\frac{1}{\mu_i} + \frac{1}{\mu_j}\right) R_{ij} + \lambda(\sigma)\left[1 + \frac{1}{2}\sum_{k=1}^{N} R_{ik}\frac{w_k}{\mu_k}\right]\right.$$

$$\left.\cdot\left[1 + \frac{1}{2}\sum_{k=1}^{N} R_{kj}\frac{w_k}{\mu_k}\right]\right\} \tag{3.8}$$

where

$$\lambda(\sigma) = \frac{1}{2} + ac\sigma + bc^2\sigma^2 . \tag{3.9}$$

The application of the technique of quasilinearization [2] yields the linear system for the $(n+1)^{st}$ approximation,

$$\frac{dR_{ij}^{n+1}}{d\sigma} = c^n\left\{-\left(\frac{1}{\mu_i} + \frac{1}{\mu_j}\right) R_{ij}^n\right.$$

$$\left. + \lambda(a^n,b^n,c^n,\sigma) f_i(R^n) f_j(R^n)\right\}$$

$$+ c^n\left\{-\left(\frac{1}{\mu_i} + \frac{1}{\mu_j}\right)(R_{ij}^{n+1} - R_{ij}^n)\right. \tag{3.10}$$

$$+ \frac{1}{2}\lambda(a^n,b^n,c^n,\sigma)\ [f_i \sum_{k=1}^{N} (R_{kj}^{n+1} - R_{kj}^n)\frac{w_k}{\mu_k}$$

$$\left. + f_j \sum_{\ell=1}^{N} (R_{i\ell}^{n+1} - R_{i\ell}^n)\frac{w_\ell}{\mu_\ell}\ ]\right\}$$

$$+ (a^{n+1} - a^n)\ c^n\ (c^n\sigma) f_i(R^n) f_j(R^n)$$

$$+ (b^{n+1} - b^n)\ c^n\ (c^n\sigma)^2\ f_i(R^n) f_j(R^n) +$$

$$+ (c^{n+1} - c^n) \left\{ - \left(\frac{1}{\mu_i} + \frac{1}{\mu_j}\right) R^n_{ij} \right.$$

$$+ \lambda(a^n, b^n, c^n, \sigma) f_i(R^n) f_j(R^n)$$

$$\left. + [a^n c^n \sigma + 2b^n (c^n \sigma)^2] \cdot f_i(R^n) f_j(R^n) \right\} ,$$

$$\frac{da^{n+1}}{d\sigma} = 0, \tag{3.10}$$

$$\frac{db^{n+1}}{d\sigma} = 0,$$

$$\frac{dc^{n+1}}{d\sigma} = 0,$$

where

$$\lambda(a^n, b^n, c^n, \sigma) = \frac{1}{2} + a^n (c^n \sigma) + b^n (c^n \sigma)^2,$$

$$f_i(R^n) = 1 + \frac{1}{2} \sum_{j=1}^{N} R^n_{ij} \frac{w_j}{\mu_j}$$

The solution of Eqs. (3.10) may be represented in the form

$$R^{n+1}_{ij}(\sigma) = p_{ij}(\sigma) + \sum_{j=1}^{3} c^k h^k_{ij}(\sigma),$$

$$a^{n+1}(\sigma) = q_1(\sigma) + \sum_{k=1}^{3} c^k v^k_1(\sigma), \tag{3.11}$$

$$b^{n+1}(\sigma) = q_2(\sigma) + \sum_{k=1}^{3} c^k v^k_2(\sigma),$$

$$c^{n+1}(\sigma) = q_3(\sigma) + \sum_{k=1}^{3} c^k v_3^k(\sigma), \tag{3.11}$$

where the vector P, constituted of elements $p_{ij}(\sigma)$ and $q_\ell(\sigma)$, is a particular solution of (3.10) and the vectors H^k composed of elements $h_{ij}^k(\sigma)$ and $v_\ell^k(\sigma)$, are three independent solutions of the homogeneous form of (5), for $k = 1, 2, 3$. We choose the initial conditions $P(0)$ identically zero, and $H^k(0)$ having all of its elements zero except for that component which corresponds to v^k, $k = 1, 2, 3$. The choice of initial conditions allows us to identify the multipliers c^k (not to be confused as powers of c) as

$$\begin{aligned} a &= a(0) = c^1 , \\ b &= b(0) = c^2 , \\ c &= c(0) = c^3 , \end{aligned} \tag{3.12}$$

We seek the three missing initial values (3.12).

Let us make the conversion from measurements of $r_{ij}(c)$ to measurements of $R_{ij}(1)$ by setting

$$\beta_{ij} = 4\mu_i b_{ij} . \tag{3.13}$$

Then we write the expression to be minimized as

$$S = \sum_{i,j} \left\{ R_{ij}^{n+1}(1) - \beta_{ij} \right\}^2 . \tag{3.14}$$

This expression is a minimum when the following requirements are met:

$$\frac{\partial S}{\partial a} = 0, \quad \frac{\partial S}{\partial b} = 0, \quad \frac{\partial S}{\partial c} = 0 . \tag{3.15}$$

By means of (3.12), these conditions are equivalent to

$$\frac{\partial S}{\partial c^1} = 0, \quad \frac{\partial S}{\partial c^2} = 0, \quad \frac{\partial S}{\partial c^3} = 0 . \tag{3.16}$$

We replace $R_{ij}^{n+1}(1)$ in (3.14) by its representation (3.11). Then Eqs. (3.16) lead us to a third order system of linear algebraic equations of the form

$$AX = B , \tag{3.17}$$

where the elements of the matrix A and the vector B are, respectively,

$$A_{ij} = \sum_{m,n} h_{mn}^i(1)\, h_{mn}^j(1) , \tag{3.18}$$

$$B_i = \sum_{m,n} h_{mn}^i(1)\, [\beta_{mn} - p_{mn}(1)] , \tag{3.19}$$

and the solution vector X has as its components the multipliers c^1, c^2, c^3. In this way we obtain the current approximation to the parameters a and b in the albedo function, and the thickness of the slab, c. To begin the calculations, we produce an initial approximation by integrating the system of nonlinear differential equations (3.8) with $R(0) = 0$, and using estimated values of the parameters. Several iterations of the method are usually sufficient to attain convergence, if convergence takes place at all.

3-5 NUMERICAL EXPERIMENTS I: MANY ACCURATE OBSERVATIONS

Some of the observations $\{\beta_{ij}\} \cong \{R_{ij}(1)\}$ are plotted in Fig. 3-3.

Several types of numerical experiments are carried out. In the first class of experiments, 49 perfectly accurate (to about 8 decimal figures) observations are used to determine the quantities a, b, and c. The 49 observations correspond to measurements for 7 outgoing angles for each of 7 incident directions, as listed in Table 2-1, Chapter 2. In one of the trials, the initial approximation is generated with the guesses $a = 2.2$ (+10% in error), $b = -1.8$ (+10% in error), and $c = 1.5$ (+50% in error). After four iterations, our estimates are decidedly better: $a = 1.99895$ (-.005% in error), $b = -1.99824$ (+.014% in error) and $c = 1.004$ (+.04% in error). We repeat the experiment, with one change: our initial estimate of the thickness is 0.5, only one-half of the correct value. This time the solution diverges and the procedure fails.

Figure 3-4a illustrates the rapid rate of convergence to the correct solution for the albedo function $\lambda(\tau)$, for the successful trial. The initial approximation is designated in the figure by the numeral 0, the first approximation by 1. The fourth approximation coincides with the true solution. Figure 3-4b shows how the initial approximation to the function $R_{ij}(c)$ for incident direction cosine 0.5 deviates from the observed values as indicated by the curve labelled "True". The first approximation lies very close to the correct values, and the fourth approximation is graphically identical with the correct solution.

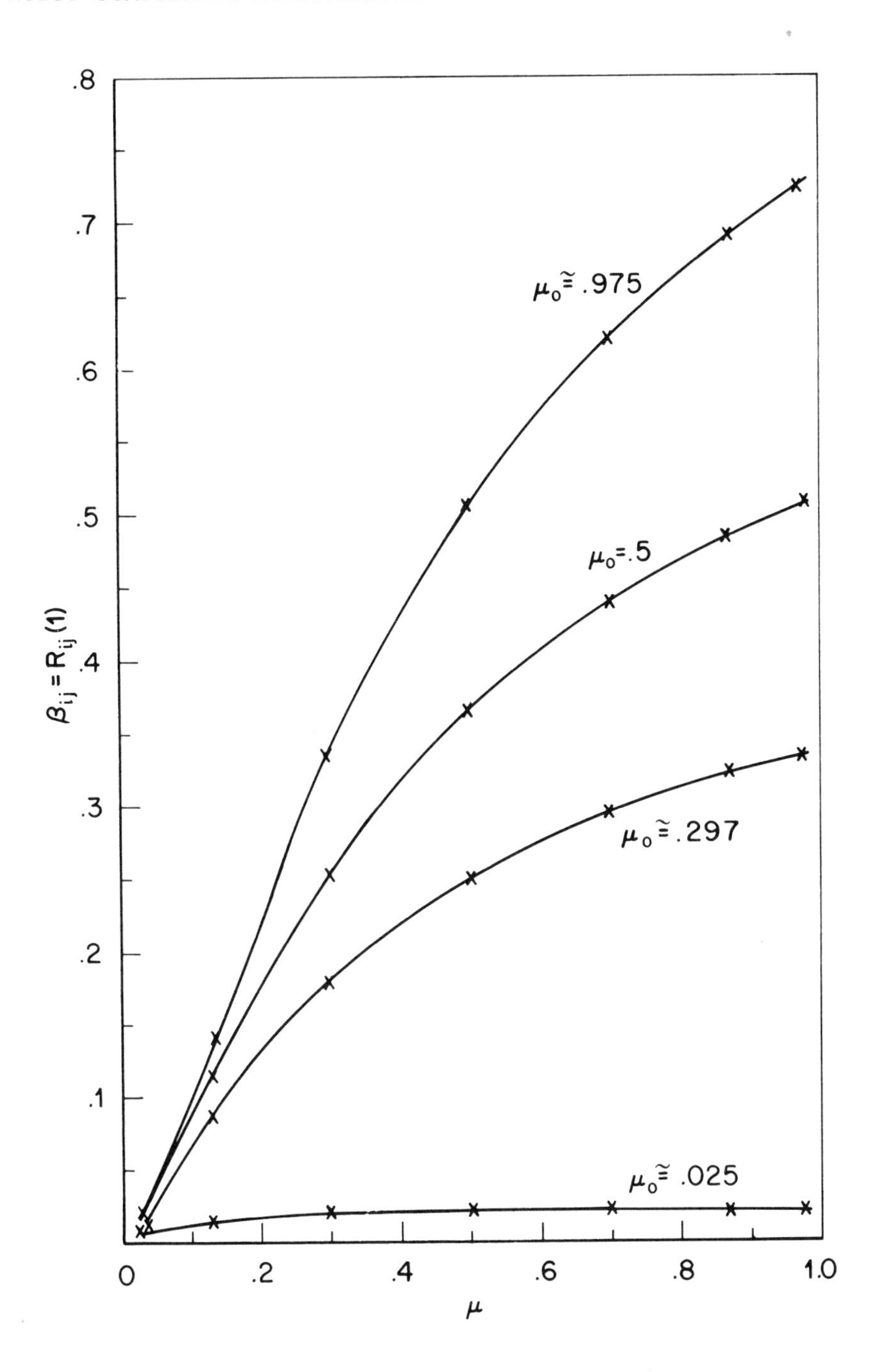

Figure 3-3 Some of the Observations $\{\beta_{ij}\}$

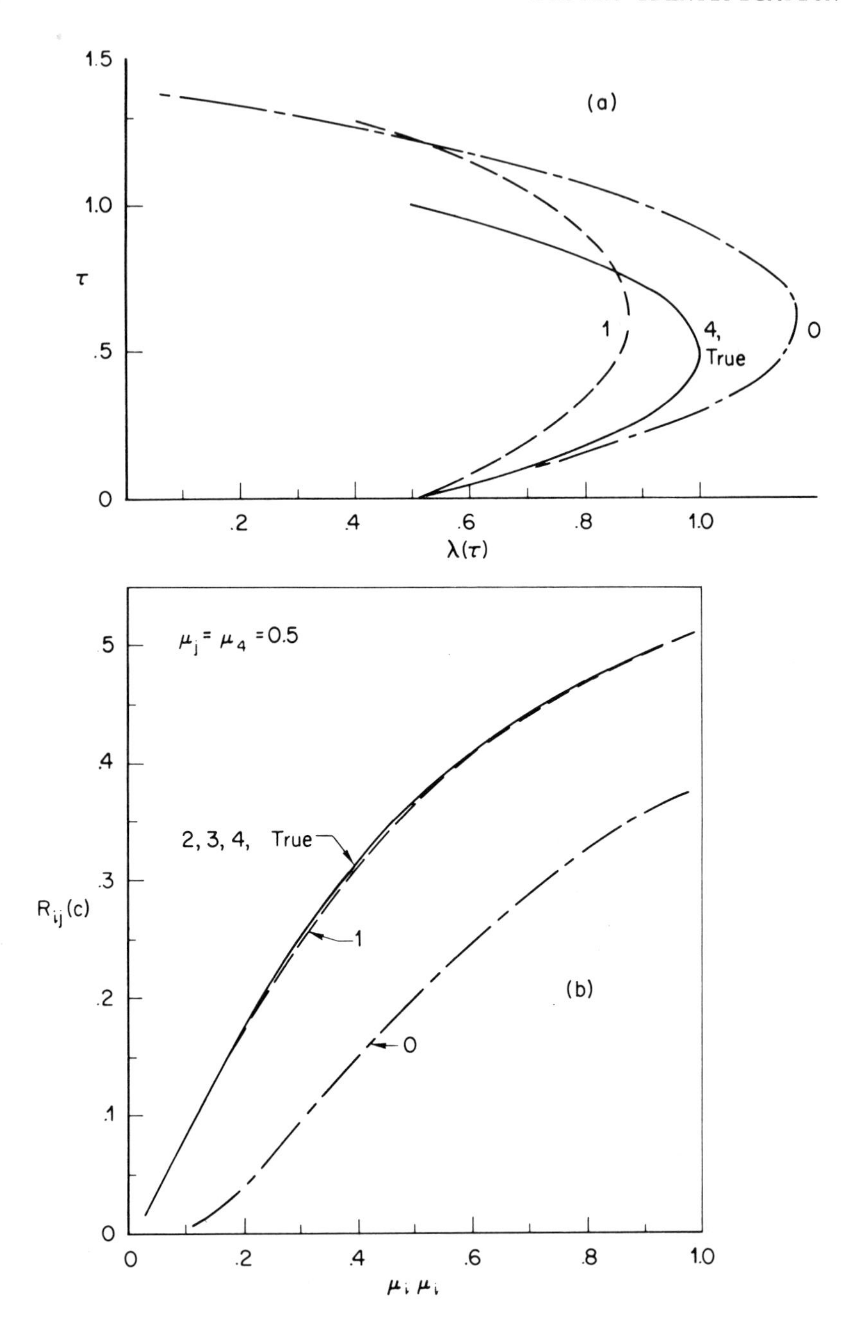

Figure 3-4 (a) Successive Approximations of the Albedo Function
(b) Successive Approximations of the Function $R_{ij}(c)$.

3-6 NUMERICAL EXPERIMENTS II: EFFECT OF ANGLE OF INCIDENCE

In a second series of experiments, the incident angle is held fixed and accurate observations are made of the outgoing radiation in seven directions. The incident direction is varied from one trial to the next in order to study the effect of the position of the source. The initial approximation used in each trial is the same, the correct solution. Due to a possible lack of information in the observations for a given trial, the successive approximations may drift away from the correct solution and converge to another. Several iterations are carried out in each run. The results of the seven runs with each of seven angles are given in Table 3-1. The incident angle is given in degrees, and the fourth approximations to the constants a, b, and the thickness c are tabulated.

Table 3-1 indicates that the results are very good, no matter what the incident angle is. Examination of the computer output shows that convergence has occurred, in each trial, to about four significant figures. Angles 13° through 72.7^{o} give nearly perfect values of the constants. Angles 82.6^{o} and 88.5^{o}, close to grazing incidence, give values which are only slightly poorer, 0.1% to 0.2% off.

3-7 NUMERICAL EXPERIMENTS III: EFFECT OF NOISY OBSERVATIONS

In this study, errors of different kinds and amounts are introduced into the observations, and the results of the determination of parameters are compared with the results of

Table 3-1

Numerical Results with Data From Various Input Directions

Trial	Incident Angle	a	b	c
1	88.5°	2.00231	-2.00456	0.999262
2	82.6°	2.00206	-2.00351	0.999361
3	72.7°	2.00032	-2.00048	0.999933
4	60.0°	2.00072	-1.99952	1.00007
5	45.3°	1.99899	-1.99841	1.00021
6	29.5°	2.00029	-2.00040	0.999972
7	13.0°	1.99962	-1.99937	1.00009
Correct Values		2.0	-2.0	1.0

Experiments I and II in which no errors were present. Errors are given in percentages with plus or minus signs. The errors in a given trial are either of equal magnitude, or they occur in a Gaussian distribution. Let $t_1, t_2, \ldots, t_7$ be seven true measurements of R. When we speak of noisy observations of $\pm$ 5% equal magnitude errors, we mean that the noisy observations are

$$n_1 = (1 + .05)t_1 ,$$

$$n_2 = (1 - .05)t_2 ,$$

$$\ldots .$$

$$n_7 = (1 + .05)t_7 .$$

Let $g_1, g_2, \ldots, g_7$ be seven (signed) Gaussian deviates, with standard deviation unity. Noisy observations with 5% Gaussian distribution of errors are defined to be

$$m_1 = (1 + .05g_1)t_1 ,$$

$$m_2 = (1 + .05g_2)t_2 ,$$

$$\ldots .$$

$$m_7 = (1 + .05g_7)t_7 .$$

The results of numerical experiments with noisy observations, with one or seven angles of incidence, are presented in Table 3-2. Clearly , the accuracy of the estimation of the three constants is in proportion to the accuracy of the observations. In contrast to the trials with perfect measurements, experiments using noisy observations are more successful when there is an abundance of data, and when the data are limited, these experiments show the effect of the incident direction. Errors with Gaussian errors give poorer results, which may be due to the particular set of 7 or 49 Gaussian deviates chosen arbitrarily from a book of random numbers [1].

3-8 NUMERICAL EXPERIMENTS IV: EFFECT OF CRITERION

This series of experiments is intended to investigate the effect of using a minimax criterion rather than a least squares condition for the determination of the unknown parameters, a, b and c. The condition requires that the constants be chosen to minimize the maximum of the absolute

Table 3-2

NUMERICAL RESULTS WITH ERRORS IN OBSERVATIONS

Incident Angle	± 1% Equal Mag. Error			± 2% Equal Mag. Error			± 5% Equal Mag. Error		
	a	b	c	a	b	c	a	b	c
88.5°	1.89	-1.80	1.05	1.79	-1.64	1.09	1.5	-1.2	1.3
82.6°	1.99	-1.96	1.013	1.975	-1.93	1.027	1.92	-1.79	1.07
72.7°	1.96	-1.93	1.016	1.92	-1.85	1.03	1.78	-1.61	1.09
60.0°	1.95	-1.91	1.016	1.89	-1.82	1.03	1.69	-1.51	1.10
45.3°	1.94	-1.90	1.016	1.87	-1.79	1.03	1.65	-1.47	1.10
29.5°	1.93	-1.89	1.016	1.86	-1.78	1.03	1.63	-1.44	1.10
13.0°	1.93	-1.89	1.016	1.86	-1.78	1.03	1.62	-1.43	1.10
All 7	1.99	-1.98	1.003	1.96	-1.94	1.009	1.91	-1.85	1.02

Incident Angle	1% Gaussian Error			2% Gaussian Error		
	a	b	c	a	b	c
88.5°	1.46	-1.2	1.22	—	—	—
82.6°	1.64	-1.41	1.15	—	—	—
72.7°	1.71	-1.55	1.09	—	—	—
60.0°	1.75	-1.63	1.06	1.54	-1.33	1.13
45.3°	1.77	-1.67	1.05	—	—	—
29.5°	1.78	-1.68	1.05	—	—	—
13.0°	1.79	-1.69	1.04	—	—	—
All 7	1.95	-1.93	1.011	—	—	—

value of the difference between $R_{ij}^{n+1}(1)$ and β_{ij}, where $R_{ij}^{n+1}(1)$ is the solution of (3.10). This is formulated as a linear programming problem in which we have the linear inequalities [2],

$$\pm \frac{1}{\beta_{ij}} \left\{ p_{ij}(1) + \sum_{k=1}^{3} c^k h_{ij}^k(1) - \beta_{ij} \right\} \leq \varepsilon_{ij} ,$$

$$\varepsilon_{ij} \leq \varepsilon , \tag{3.20}$$

where the subscripts take on the values appropriate to the trial under consideration. A standard linear programming code [2, 3] is used to determine the constants c^k, ε_{ij}, and the maximum deviation ε. Two numerical experiments are carried out, one with $\pm$ 2% equal magnitude errors in the observations, the other with 2% Gaussian errors. The incident angle is 60°. The results are given in Table 3-3, where we show the values of the two constants in the albedo functions, a and b, the thickness c, and the maximum deviation ε, for each approximation. The results for the case where the errors are all of the same relative size are excellent. The trial using Gaussian errors yields constants which are not quite as good, yet these results are surprisingly better than one might expect.

3-9 NUMERICAL EXPERIMENTS V: CONSTRUCTION OF MODEL ATMOSPHERES

Suppose that we desire to construct a model atmosphere with the optical property that whenever light is

Table 3-3

Numerical Results Using Minimax Criterion

Type of Errors	Approximation	a	b	c	Maximum Deviation
$\pm$ 2% equal magnitude	0	2.00000	-2.00000	1.00000	———
	1	1.99948	-1.99961	1.00001	.0200000
	2	1.99948	-1.99959	1.00001	.0200000
	3	1.99948	-1.99960	1.00001	.0200000
2% Gaussian	0	2.00000	-2.00000	1.00000	———
	1	1.76462	-1.67267	1.03357	.0294158
	2	1.76265	-1.67487	1.03841	.0293736
	3	1.76279	-1.67484	1.03852	.0293722

incident at angles near the normal, the distribution of diffusely reflected light is greatest close to 90° from the normal. We require that the optical thickness c be about 1.0, and the albedo profile is to be parabolic,

$$\lambda(\tau) = \frac{1}{2} + a\tau + b\tau^2 , \tag{3.21}$$

where the constants a and b are to be suitably chosen. The albedo should not be greater than unity.

The reflection pattern, for an incident angle of 13°, is to have the form indicated by the seven x's in Fig. 3-5a. The units are given relative to an incident net flux of π. As our initial estimate, we believe that the slab should have thickness one, and that the parameters be $a = 2$, and $b = -2$. Then the albedo has the form given in Fig. 3-5b by the curve labelled "Initial", the horizontal line at $\tau = 1$ indicating the upper surface c. The reflection function has the form given in Fig. 3-5a by the dots, whose values are much too low in the region 80°-90°. How should the optical design of this slab be modified for better agreement with the requirements? The answer is not at all obvious.

We carry out a numerical experiment in which a better model is to be found, which makes the sum of the squares of the deviations from the desired values a minimum. The condition is to minimize the sum S,

$$S = \sum_{i=1}^{7} (d_i - r_{ik})^2 , \tag{3.22}$$

where d_i is the desired value of the reflection function for output angle arc cosine μ_i, and r_{ik} is the solution of the differential equations for the r function, and $k = 7$, corresponding to input angle of 13°. This problem is mathematically the same as the earlier inverse problems of this chapter. The method of solution is also similar, and after five iterations and 3 minutes of computing time, we obtain the solution $a = 1.383$, $b = -1.140$, and $c = 1.117$. The albedo function is shown in Fig. 3-5b by the curve labelled "Least squares", the reflection function is indicated by the circled dots in Fig. 3-5a. A smooth curve is drawn between the dots, showing the probable continuous distribution. This curve is in better agreement with the requirements at 83° and 88.5°.

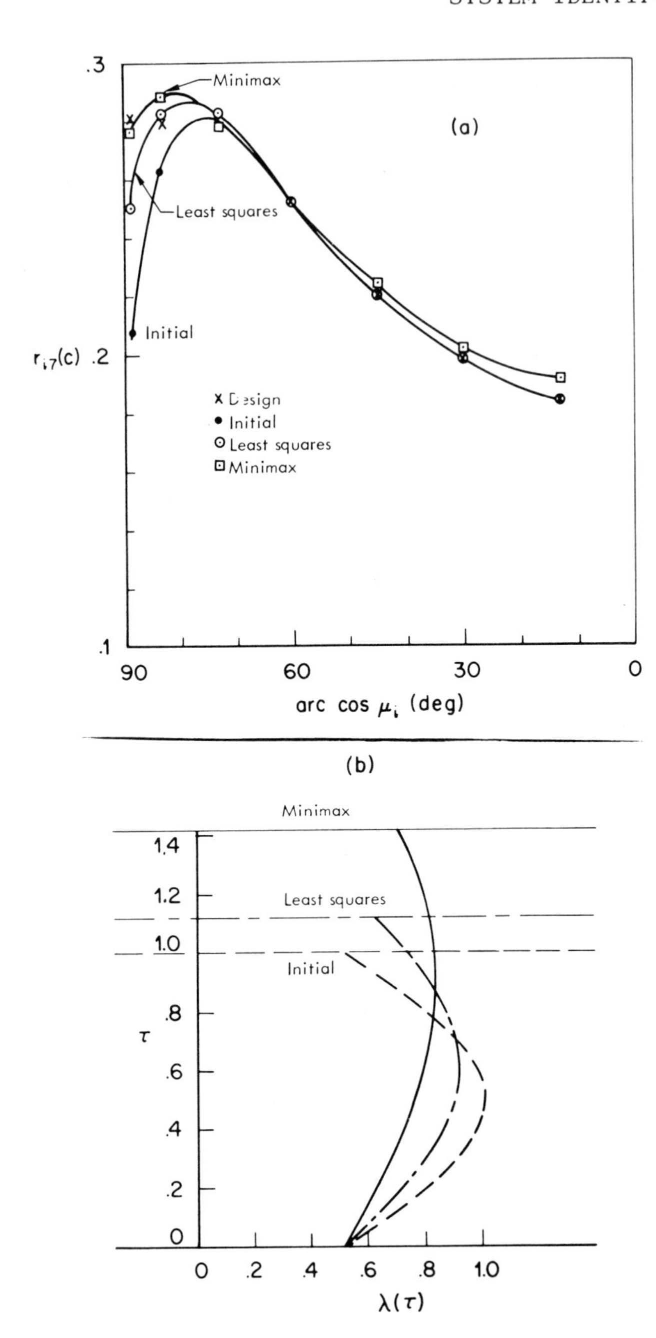

Figure 3-5 Several Model Atmospheres

(a) The diffusely reflected intensity with incident angle 13°.

(b) The albedo function.

We perform another experiment in which the criterion is to minimize the maximum deviation. This condition is given by Eqs. (3.20), where $\beta_{ij} = 4\mu_i d_i$ and $j = 7$. After five iterations, the solution is $a = 0.744$, $b = -0.415$, and optical thickness $c = 1.431$. The albedo has the form represented in Fig. 3-5b by the curve labelled "Minimax", and the reflection function is indicated by dots within squares in Fig. 3-5a. The reflection function for this slab is in very good agreement with the requirements.

Other possible approaches to problems of design include dynamic programming and invariant imbedding.

REFERENCES

1. The Rand Corporation, A Million Random Digits with 100,000 Normal Deviates, The Free Press, Glencoe, Illinois, 1955.

2. Dantzig, G., Linear Programming and Extensions, Princeton University Press, Princeton, New Jersey, 1963.

3. Clasen, R. J., "RS MSUB - Linear Programming Subroutine, FORTRAN IV Coded," RAND 7044 Library Routine W009, The Rand Corporation, Santa Monica, 1964.

CHAPTER 4

INVERSE PROBLEMS IN RADIATIVE TRANSFER: ANISOTROPIC SCATTERING

4-1 INTRODUCTION

Inverse problems in radiative transfer for a medium with anisotropic scattering [1, 2] may be treated in a manner similar to that used in the isotropic case. We consider a plane parallel slab of finite optical thickness τ_1. For simplicity let us suppose that both the albedo λ and the anisotropic phase function are independent of optical height. Let us take the phase function to be

$$p(\theta) = 1 + a \cos \theta, \quad \frac{1}{4\pi} \int_{\Omega} p \, d\Omega = 1, \tag{4.1}$$

where θ is the scattering angle, and a is a parameter of the medium and is to be determined on the basis of measurements of the diffusely reflected radiation. An integration over solid angle gives the normalization condition in (4.1). Let us consider the case in which $a = 1$. This approximately corresponds to the forward phase diagram of Saturn [3]. It may be noted that Horak considers inverse problems in planetary atmospheres in [3].

Parallel rays of light of net flux π per unit area normal to the direction of propagation are incident on the upper surface of the slab. The direction of the rays is characterized by μ_0 $(0 < \mu_0 \leq 1)$, the cosine of the polar angle measured from the downward normal, and by the azimuth angle ϕ_0. The phase function may be written as a function of polar and azimuth angles of incidence and scattering, $p(\mu, \phi; \mu_0, \phi_0)$. The lower surface of the medium is a perfect absorber.

Let the diffusely reflected intensity in the direction specified by (μ, ϕ) be $r(\tau_1, \mu, \phi; \mu_0, \phi_0)$, where ϕ is the azimuth angle $(0 \leq \phi \leq 2\pi)$. Measurements of r, the set $\{b\}$, are made and we wish to determine the value of a for the slab.

4-2 THE S FUNCTION

Let us define a function S, which is related to the diffusely reflected intensity function r, by the formula

$$r(\tau_1,\mu,\phi; \mu_0,\phi_0) = \frac{S(\tau_1,\mu,\phi; \mu_0,\phi_0)}{4\pi} . \tag{4.2}$$

We wish to derive a differential-integral equation for the S function by the method of invariant imbedding (see also [1, 2]).

Let us define another function ρ,

$$\rho(\tau_1,\mu,\phi;\ \mu_0,\phi_0) = \frac{\mu}{\mu_0}\ \frac{r(\tau_1,\mu,\phi;\ \mu_0,\phi_0)}{\pi} \tag{4.3}$$

which is the reflected radiation per unit horizontal area produced by a unit input of radiation on a unit of area in the top surface. Now we add a thin layer of thickness Δ to the top of the slab of thickness τ_1. The ρ function for this slab may be expressed in the form,

$$\begin{aligned} \rho(\tau_1+\Delta,\mu,\phi;\ \mu_0,\phi_0) &= \rho(\tau_1,\mu,\phi;\ \mu_0,\phi_0) \\ &- \Delta\left(\frac{1}{\mu}+\frac{1}{\mu_0}\right)\rho(\tau_1,\mu,\phi;\ \mu_0,\phi_0) \\ &+ \frac{\Delta}{\mu_0}\ \frac{\lambda}{4\pi}\ p(\mu,\phi;\ -\mu_0,\phi_0) \\ &+ \int_0^{2\pi}\int_0^1 \rho(\tau_1,\mu',\phi';\ \mu_0,\phi_0)\ \frac{\Delta}{\mu'}\ \frac{\lambda}{4\pi} \\ &\cdot\ p(\mu,\phi;\ \mu',\phi')\ d\mu'\ d\phi' \\ &+ \int_0^{2\pi}\int_0^1 \frac{\Delta}{\mu}\ \frac{\lambda}{4\pi}\ p(-\mu_0',\phi_0';\ -\mu_0,\phi_0) \\ &\cdot\ \rho(\tau_1,\mu,\phi;\ \mu_0',\phi_0')\ d\mu_0'\ d\phi_0'\ + \end{aligned} \tag{4.4}$$

$$+\int_0^{2\pi}\int_0^1 \rho(\tau_1,\mu',\phi';\ \mu_0,\phi_0)\,\frac{\Delta}{\mu'}\,\frac{\lambda}{4\pi}$$

$$\cdot\ p(-\mu_0',\phi_0';\ \mu',\phi')\ d\mu'\ d\phi' \qquad (4.4)$$

$$\cdot \int_0^{2\pi}\int_0^1 \rho(\tau_1,\mu,\phi;\ \mu_0',\phi_0')\ d\mu_0'\ d\phi_0' + o(\Delta)\ .$$

The terms on the right-hand side of the equation represent the following processes: (1) no interaction in Δ, (2) absorption in Δ of the incoming and outgoing rays, (3) a single scattering in Δ, (4) multiple scattering in the slab of thickness τ_1 followed by an interaction in Δ, (5) interaction in Δ followed by multiple scattering in the slab below, (6) multiple scattering in the slab of thickness τ_1 followed by an interaction in Δ, followed by multiple scattering in the slab of thickness Δ, and (7) $o(\Delta)$ represents other processes which involve Δ^2, or higher powers of Δ.

Now the relation between S and ρ is

$$\rho(\tau_1,\mu,\phi;\ \mu_0,\phi_0) = \frac{S(\tau_1,\mu,\phi;\ \mu_0,\phi_0)}{4\pi\mu_0}$$

so that when we substitute this expression into (4.4), we find that S satisfies the equation

$$S(\tau_1+\Delta,\mu,\phi;\ \mu_0\phi_0) = S(\tau_1,\mu,\phi;\ \mu_0,\phi_0) \qquad (4.6)$$

$$-\ \Delta\left(\frac{1}{\mu}+\frac{1}{\phi_0}\right) S(\tau_1,\mu,\mu;\ \mu_0,\phi_0)\ +$$

$$
\begin{aligned}
&+ \Delta\, \lambda\, p(\mu,\phi;\, -\mu_0,\phi_0) \\
&+ \frac{\Delta\lambda}{4\pi}\int_0^{2\pi}\int_0^1 S(\tau_1,\mu',\phi';\, \mu_0,\phi_0) \\
&\cdot\, p(\mu,\phi;\, \mu'\phi')\,\frac{d\mu'}{\mu'}\,d\phi' \\
&+ \frac{\Delta\lambda}{4\pi}\int_0^{2\pi}\int_0^1 p(-\mu_0',\phi_0';\, -\mu_0,\phi_0) \\
&\cdot\, S(t_1,\mu,\phi;\, \mu_0',\phi_0')\,\frac{d\mu_0'}{\mu_0'}\,d\phi_0' \\
&+ \frac{\Delta\lambda}{(4\pi)^2}\int_0^{2\pi}\int_0^1 S(\tau_1,\mu',\phi';\, \mu_0,\phi_0) \\
&\cdot\, p(-\mu_0',\phi_0';\, \mu',\phi')\,\frac{d\mu'}{\mu'}\,d\phi' \\
&\cdot\int_0^{2\pi}\int_0^1 S(\tau_1,\mu,\phi;\, \mu_0',\phi_0')\,\frac{d\mu_0'}{\mu'}\,d\phi_0' + o(\Delta)\ .
\end{aligned}
\tag{4.6}
$$

We expand the left-hand side of Eq. (4.6) in powers of Δ,

$$
\begin{aligned}
S(\tau_1+\Delta,\mu,\phi;\, \mu_0,\phi_0) &= S(\tau_1,\mu,\phi;\, \mu_0,\phi_0) \\
&+ \frac{\partial S(\tau_1,\mu,\phi;\, \mu_0,\phi_0)}{\partial \tau_1} + o(\Delta)\ .
\end{aligned}
\tag{4.7}
$$

We let $\Delta \to 0$ and we obtain the desired integro-differential equation

$$\frac{\partial S(\tau_1,\mu,\phi;\ \mu_0,\phi_0)}{\partial \tau_1} + \left(\frac{1}{\mu} + \frac{1}{\mu_0}\right) S = \lambda\{p(\mu,\phi;\ -\mu_0,\phi_0)$$

$$+ \frac{1}{4\pi}\int_0^{2\pi}\int_0^1 S(\tau_1,\mu',\phi';\ \mu_0,\phi_0)$$

$$\cdot\ p(\mu,\phi;\ \mu',\phi')\ \frac{d\mu'}{\mu'}\ d\phi'$$

$$+ \frac{1}{4\pi}\int_0^{2\pi}\int_0^1 S(\tau_1,\mu,\phi;\ \mu_0',\phi_0') \tag{4.8}$$

$$\cdot\ p(-\mu_0',\phi_0';\ -\mu_0,\phi_0)\ \frac{d\mu_0'}{\mu'}\ d\phi_0'$$

$$+ \frac{1}{(4\pi)^2}\int_0^{2\pi}\int_0^1 S(\tau_1,\mu',\phi';\ \mu_0,\phi_0)$$

$$\cdot\ p(-\mu_0',\phi';\ \mu',\phi')\ \frac{d\mu'}{\mu'}\ d\phi'$$

$$\cdot\int_0^{2\pi}\int_0^1 S(\tau_1,\mu,\phi;\ \mu_0',\phi_0')\ \frac{d\mu_0'}{\mu_0'}\ d\phi_0'\} \ .$$

A simplification arises if it is assumed that the phase function may be expanded in the Fourier series

$$p(\mu,\phi,\mu_0,\phi_0) = \sum_{m=0}^{M} c_m P_m(\cos\theta) \ , \tag{4.9}$$

where $P_m(x)$ is the Legendre polynomial of degree m. The angular dependence of expansion (4.9) may be decomposed into polar and azimuth factors by the use of the addition rule of Legendre functions. Then Eq. (4.9) becomes

$$p(\mu,\phi,\mu_0,\phi_0) = \sum_{m=0}^{M} (2-\delta_{0m})$$

$$\cdot \sum_{i=m}^{M} c_i \frac{(i-m)!}{(i+m)!} P_i^m(\mu) P_i^m(\mu_0) \cos m(\phi-\phi_0) . \tag{4.10}$$

The function $P_i^m(x)$ is the associated Legendre function of degree i, order m. Noting the form of this equation, we expand the S function in a similar manner,

$$S(\tau_1,\mu,\phi,\mu_0,\phi_0) = \sum_{m=0}^{M} S^{(m)}(\tau_1,\mu,\mu_0) \cos m(\phi-\phi_0) . \tag{4.11}$$

Substitution of (4.11) in (4.8) leads to the equations for the the Fourier components of S,

$$\frac{\partial S^{(m)}}{\partial \tau_1} + (\frac{1}{\mu} + \frac{1}{\mu_0}) S^{(m)} = \lambda(2-\delta_{0m}) \sum_{i=m}^{M} (-1)^{i+m}$$

$$\cdot c_i \frac{(i-m)!}{(i+m)!} \psi_i^m(\mu) \psi_i^m(\mu_0) \tag{4.12}$$

where

$$\psi_i^m(\mu) = P_{ii}^m(\mu) + \frac{(-1)^{i+m}}{2(2-\delta_{0m})} \int_0^1 S^{(m)}(\tau_1,\mu,\mu') P_i^m(\mu') \frac{d\mu'}{\mu'} , \tag{4.13}$$

for $m = 0, 1, 2, \ldots, M$. The functions $S^{(m)}(\tau_1,\mu,\mu_0)$ possess the symmetry property

$$S^{(m)}(\tau_1,\mu,\mu_0) = S^{(m)}(\tau_1,\mu_0,\mu) \ . \tag{4.14}$$

The initial conditions are

$$S^{(m)}(0,\mu,\mu_0) = 0 \ . \tag{4.15}$$

By the use of Gaussian quadrature on the interval (0,1), the integrals (4.13) are replaced by sums. Also, the function $S^{(m)}(\tau_1,\mu,\mu_0)$ is replaced by a function of one independent variable, $S_{ij}^{(m)}(\tau_1)$, where the angles are discretized such that $\mu_0 \to \mu_j$, and $\mu \to \mu_i$, $i,j = 1, 2, \ldots, N$. Then we have the approximate system,

$$\frac{dS_{ij}^{(m)}(\tau_1)}{d\tau_1} + \left(\frac{1}{\mu_i} + \frac{1}{\mu_j}\right) S_{ij}^{(m)} \tag{4.16}$$

$$= \lambda(2-\delta_{0m}) \sum_{k=m}^{M} (-1)^{k+m} \frac{(k-m)!}{(k+m)!} c_k \psi_{ki}^m \psi_{kj}^m \ ,$$

$$(m = 0, 1, \ldots, M; \quad i = 1, 2, \ldots, N;$$

$$j = 1, 2, \ldots, N) \ ,$$

where

$$\psi_{k\ell}^m = P_k^m(\mu_\ell) + \frac{(-1)^{k+m}}{2(2-\delta_{0m})} \sum_{j=1}^{N} S_{\ell j}^{(m)} P_k^m(\mu_j) \frac{w_j}{\mu_j} \ . \tag{4.17}$$

The discrete cosines μ_j are the roots of the shifted Legendre polynomial of degree N, $P_N^*(x)$, and the quantities w_j are the corresponding weights. The initial conditions are

$$S_{ij}^{(m)}(0) = 0 . \tag{4.18}$$

The solution of this initial value integration problem for a system of ordinary differential equations (4.16) is approximately equal to the solution of the original integro-differential system.

4-3 AN INVERSE PROBLEM

Consider the case in which the slab is of thickness $\tau_1 = 0.2$, the albedo is $\lambda = 1$, and we choose $N = 7$ for the quadrature. For the phase function

$$p = 1 + a \cos \theta, \tag{4.19}$$

the parameters are

$$M = 1, \; c_0 = 1, \; c_1 = a = 1 . \tag{4.20}$$

For a numerical experiment, we take Eqs. (4.16) and integrate from $\tau_1 = 0$ with initial conditions (4.18) to $\tau_1 = 0.2$, using an integration grid size of $\Delta\tau_1 = 0.01$. Using (4.11) and (4.2), we produce

$$b_{ijk} \cong r(0.2,\mu_i,\phi_k;\ \mu_j,\phi_0) \tag{4.21}$$

for $\phi_0 = 0$, and $\phi_k = 0^\circ, 30^\circ, 60^\circ, \ldots, 180^\circ$ as $k = 1, 2, \ldots, 7$. The set $\{b_{ijk}\}$ represents our measurements of the diffuse reflection field, from which we hope to estimate the unknown parameter a. The condition shall be to minimize the sum of squares of deviations,

$$\sum_{i,j,k} [r(0.2,\mu_i,\phi_k;\ \mu_j,\phi_0) - b_{ijk}]^2$$

where the function r is the solution of the Eqs. (4.16) - (4.18) using (4.2) and (4.11). The measurements for $\phi = 0^\circ$ and for $\phi = 180^\circ$ are shown in Fig. 4-1, when $\mu_0 = 0.5$, $\phi_0 = 0$. These data were produced numerically with the use of Eqs. (4.2), (4.11) - (4.15).

4-4 METHOD OF SOLUTION

This problem may be solved by successive approximations using quasilinearization. Let us write the function $S_{ij}^{(m)}$ as S_{mij}, and similarly $\psi_{ki}^{m} \to \psi_{mki}$, $P_k^m(\mu_\ell) \to P_{mk\ell}$. The linear equations for the $(n+1)^{st}$ approximation are

$$\frac{dS_{mij}^{n+1}}{d\tau_1} = -\left(\frac{1}{\mu_i} + \frac{1}{\mu_j}\right) S_{mij}^{n}$$
$$+ \lambda(2-\delta_{0m}) \sum_{k=m}^{1} (-1)^{k+m} \frac{(k-m)!}{(k+m)!} c_k^n \psi_{mki}\psi_{mkj} + \tag{4.23}$$

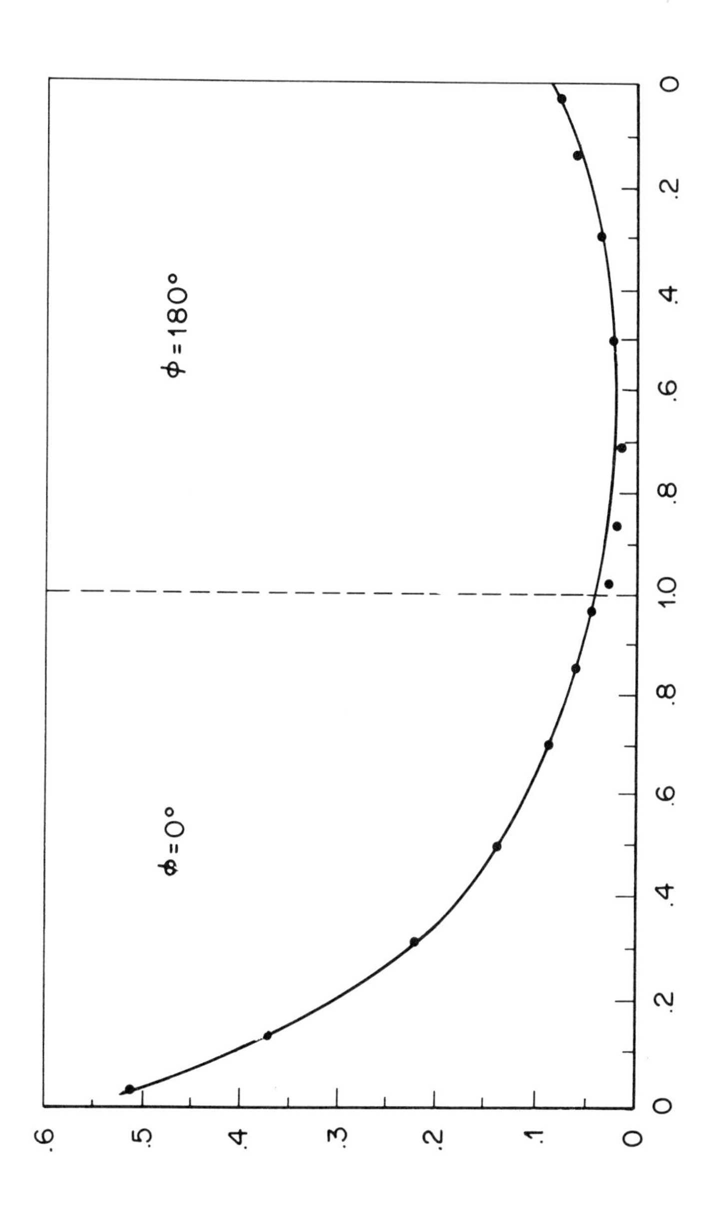

Figure 4-1 Fourteen Observations of the Diffusely Reflected Intensity $r(\mu,\phi;\mu_0,\phi_0)$, $\mu_0 = 0.5$, $\phi_0 = 0^{\circ}$, Phase Function $p(\theta) = 1 + \cos\theta$.

$$+ (S^{n+1}_{mij} - S^{n}_{mij})(-1)(\frac{1}{\mu_i} + \frac{1}{\mu_j})$$

$$+ \lambda(2-\delta_{0m}) \sum_{\ell=1}^{N} [(S^{n+1}_{mj\ell} - S^{n}_{mj\ell})\Phi_{mi\ell} \tag{4.23}$$

$$+ (S^{n+1}_{mi\ell} - S^{n}_{mi\ell})\Phi_{mj\ell}]$$

$$+ \lambda(2-\delta_{0m})(c_1^{n+1} - c_1^{n})(-1)^{1+m} \frac{(1-m)!}{(1+m)!} \psi_{m1i}\psi_{m1j} ,$$

$$(m = 0, 1; i = 1, 2, \ldots, 7; j = 1, 2, \ldots, 7),$$

$$\frac{da^{n+1}}{d\tau_1} = 0 \quad \text{for} \quad a = c_1 , \tag{4.24}$$

where

$$\Phi_{mi\ell} = \sum_{k=m}^{1} \frac{(-1)^{k+m}}{2(2-\delta_{0m})} P_{mk\ell} \frac{w_\ell}{\mu_\ell} (-1)^{k+m} \frac{(k-m)!}{(k+m)!} c_k^n \psi_{mki} \tag{4.25}$$

and

$$\psi_{mk\ell} = P_{mk\ell} + \frac{(-1)^{k+m}}{2(2-\delta_{0m})} \sum_{j=1}^{N} S^{n}_{m\ell j} P_{mkj} \frac{w_j}{\mu_j} . \tag{4.26}$$

In these equations

$$a^{n+1} = c_1^{n+1} , \ a^n = c_1^n , \ c_0^{n+1} = c_0^n = 1 . \tag{4.27}$$

The initial conditions are

$$S_{mij}^{n+1}(0) = 0 \tag{4.28}$$

and the boundary condition is

$$\frac{\partial}{\partial a^{n+1}} \left\{ \sum_{i,j,k} \left[\sum_{m=0}^{1} S_{mij}^{n+1}(0.2) \cos m\phi_k - 4\mu_i b_{ijk} \right]^2 \right\} = 0. \tag{4.29}$$

Let us represent the $(n+1)^{st}$ approximation of S as a linear combination of a particular solution and a homogeneous solution

$$S_{mij}^{n+1}(\tau_1) = p_{mij}(\tau_1) + a\, h_{mij}(\tau_1) . \tag{4.30}$$

In terms of numerically known quantities,

$$a^{n+1} = \frac{\sum_{i,j,k} (4\mu_i b_{ijk} - p_{0ij} - p_{1ij} \cos\phi k)(h_{0ij} + h_{1ij} \cos\phi k)}{\sum_{i,j,k} [h_{0ij} + h_{1ij} \cos\phi k]^2}$$

where the functions p and h are evaluated at $\tau_1 = 0.2$, and the initial conditions for p and h are suitabley chosen.

By making use of the symmetry property of S, we need consider, not a system of $2N^2$ equations, but only $2N(N+1)/2 = N(N+1)$ equations. For $N = 7$, this means that $7.8 = 56$ equations define the particular solution, and another 56 define the homogeneous solution. Twenty-one integration grid points cover the range $0 \leq \tau_1 \leq 0.2$, with $\Delta\tau_1 = 0.01$. The storage requirements are 21 x 56 for the p solution, 21 x 56 for the h solution and 21 x 56 for S^n_{mij}. Numerical studies should prove useful in the planning and analysis of investigations of planetary atmospheres [3, 4-13], stellar radiation in the galaxy [14], and radiation fields in the sea [15-17].

4-5 NUMERICAL RESULTS

Several computational experiments are performed. In three controlled experiments without any errors in the observations, but with wrong initial estimates $(a = 0{\cdot}8, 0{\cdot}5,$ and $0{\cdot}0)$, the refined estimate $a = 0{\cdot}99999$ is obtained. Note that setting $a = 0$ implies isotropic scattering as the initial estimate.

In the second series of experiments, errors of 0·5 1, 2, 5, and 10 percent Gaussian noise are introduced into the observations of r, and estimates of the parameter a are obtained which are all less than approximately 1 percent in discrepancy. In the experiment with 10 percent error and with isotropic scattering as the initial estimate, the sequence of values of the coefficient a is as follows:

$$a = 0{\cdot}0 \,,$$

$$a = 1{\cdot}087 \,,$$

$$a = 1{\cdot}014 \,,$$

$$a = 1{\cdot}014 \,.$$

Thus, even with large errors in the measurements and with a completely wrong initial estimate of the coefficient, the estimate of the coefficient is refined in only two steps to a discrepancy of only 1·4 percent. This calculation requires less than 1·5 minutes of computing time.

REFERENCES

1. Sobolev, V. V., A Treatise on Radiative Transfer, D. Van Nostrand Co., Princeton, New Jersey, 1963.

2. Kagiwada, H., and R. Kalaba, "Exact Solution of A Family of Integral Equations of Anisotropic Scattering," J. Math. Phys., Vol. 11, pp. 1575-1578, 1970.

3. Horak, Henry G., "Diffuse Reflection by Planetary Atmospheres," Ap. J., Vol. 112, pp. 445-463, 1950.

4. Dave, J. V., "Meaning of Successive Iteration of the Auxiliary Equation in the Theory of Radiative Transfer," Astrophys. J., Vol. 140, No. 3, pp. 1292-1302, 1964.

5. Deirmendjian, D., "A Water Cloud Interpretation of Venus' Microwave Spectrum," Icarus, Vol. 3, No. 2, pp. 109-120, 1964.

6. Schilling, G. F., "Extreme Model Atmosphere of Mars," Memoires Soc. R. Sc. Liege, cinquieme serie, tome VII, pp. 448-451, 1962.

7. Van de Hulst, H. C., and W. M. Irvine, "General Report on Radiation Transfer in Planets: Scattering in Model Planetary Atmospheres," Mémoires Soc. R. Sc. Liège, cinquième série, tome VII, pp. 78-98, 1962.

8. Sekera, Z., and J. V. Dave, "Diffuse Transmission of Solar Ultraviolet Radiation in the Presence of Ozone," Astrophys. Journal, Vol. 133, pp. 210-227, 1961.

9. Lenoble, J., and Z. Sekera, "Equation of Raidative Transfer in a Planetary Spherical Atmosphere," Proc. Nat. Acad. Sci., Vol. 47, No. 3, pp. 372-378, 1961.

10. Van de Hulst, H. C., "Scattering in a Planetary Atmosphere," Astrophys. Journal, Vol. 107, pp. 220-246, 1948.

11. Chu, C. M., J. A. Leacock, J. C. Chen and S. W. Churchill, "Numerical Solutions for Multiple, Anisotropic Scattering," Electromagnetic Scattering, edited by M. Kerker, Pergamon Press, Oxford, 1963, pp. 567-582.

12. Malkevich, M. S., "Angular and Spectral Distribution of Radiation Reflected by the Earth into Space," Planet. Space Sci., Vol. 11, pp. 681-699, 1963.

13. Wernik, A., "Extinction of the Night Sky Light in an Anisotropically Scattering Atmosphere," Acta Astronomica, Vol. 12, No. 2, pp. 102-121, 1962.

14. Henyey, L. G., and J. L. Greenstein, "Diffuse Radiation in the Galaxy," Astrophys. Journal, Vol. 93, pp. 70-83, 1941.

15. Preisendorfer, R. W., "Application of Radiative Transfer Theory to Light Measurements in the Sea," L'Institut Géographique National, Paris, International Union of Geodesy and Geophysics, Monographie No. 10, June 1961.

16. Tyler, J. E., "Radiance Distribution as a Function of Depth in an Underwater Environment," Bulletin of the Scripps Institution of Oceanography, Vol. 7, No. 5, pp. 363-412, 1960.

17. Tyler, J. E., and A. Shaules, "Irradiance on a Flat Object Underwater," Applied Optics, Vol. 3, pp. 105-110, 1964.

CHAPTER 5

AN INVERSE PROBLEM IN NEUTRON TRANSPORT THEORY

5-1 INTRODUCTION

The theory of neutron transport and the theory of radiative transfer [1] are devoted to problems of determining the properties of radiation fields produced by given sources in a given medium. Inverse problems in transport theory are those in which we seek to determine the properties of the medium, given those of the indicent radiation and the radiation fields [2-4].

In this chapter, we study inverse problems in transport theory from the point of view of dynamic programming [5]. Our aim is to produce a feasible computational method for estimating the properties of the medium based upon measurements of radiation fields within the medium. The invariant imbedding approach to transport theory is sketched in Ref. 6.

For ease of exposition we consider a one-dimensional transport process. The method described here can be generalized to the vector-matrix case, and thus to the slab geometry with anisotropic scattering, to wave propagation [7], and to transmission lines.

5-2 FORMULATION

Consider the one-dimensional medium shown in Fig. 5-1.

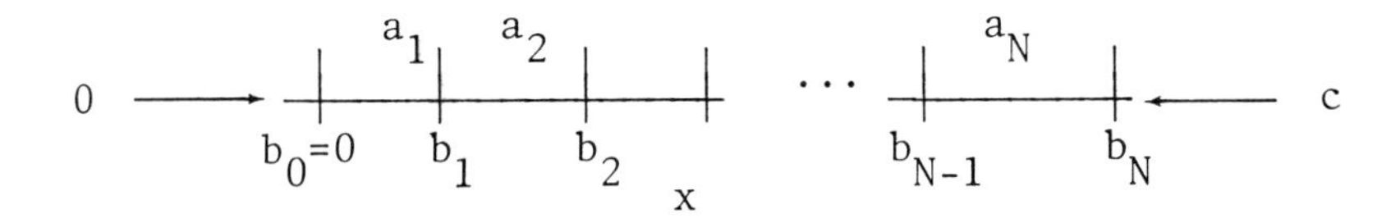

Figure 5-1 A One-dimensional Transport Process

It consists of N homogeneous sections (b_i, b_{i+1}), $i = 0, 1, 2, \ldots, N-1$. When a neutron travels through a distance Δ in the ith section, there is probability $a_i\Delta$ that it will interact with the medium. The result of an interaction is that the original neutron is absorbed and two daughter neutrons appear, one traveling in each direction. Suppose that c neutrons per unit time are incident from the right and zero neutrons per unit time from the left. We denote the average number of particles per unit time passing the point x and moving to the right by $u(x)$ and the same quantity for the leftward moving particles by $v(x)$. Suppose that measurements on the internal intensities are made at various points $x = x_i \neq b_j$; e.g.,

$$u(x_i) \cong w_i, \quad i = 1, 2, \ldots, M . \tag{5.1}$$

Our aim is to estimate the characteristics of the medium, the quantities a_i, $i = 1, 2, \ldots, N$, on the basis of these observations.

As is shown in Ref. 6, the internal intensities satisfy the differential equations

$$\begin{aligned} \dot{u} &= a_i v , \\ -\dot{v} &= a_i u , \end{aligned} \tag{5.2}$$

$$b_{i-1} \leq x \leq b_i, \quad i = 1, 2, \ldots, N,, \tag{5.3}$$

where the dot indicates differentiation with respect to x. The analytical solution is of no import, since we wish to consider this as a prototype of more complex processes for which a computational treatment is mandatory. In addition, $u(x)$ and $v(x)$ are continuous at the interfaces

$$u(b_i - 0) = u(b_i + 0) \tag{5.4}$$

$$v(b_i - 0) = v(b_i + 0), \quad i = 1, 2, \ldots, N - 1 , \tag{5.5}$$

and

$$u(0) = 0 \tag{5.6}$$

$$v(b_N) = c . \tag{5.7}$$

We wish to select the N constants $a_i, a_2, \ldots, a_N$, so as to minimize the sum of the squares of the deviations S,

$$S = \sum_{i=1}^{M} \{u(x_i) - w_i\}^2 \,. \tag{5.8}$$

5-3 DYNAMIC PROGRAMMING

Let us suppose that the functions u and v are subject to the conditions of Section 5-2 and

$$u(b_k) = c_1 \tag{5.9}$$

$$v(b_k) = c_2 \,. \tag{5.10}$$

In addition we write

$$f_K(c_1,c_2) = \min \sum_{i=1}^{M_K} \{u(x_i) - w_i\}^2 \,, \tag{5.11}$$

where the minimization is over the absorption coefficients $a_1, a_2, \ldots, a_K$. The number of observations on the first K intervals is M_K. We view K as a parameter taking on the values $1, 2, \ldots$, and c_1 and c_2 are also viewed as variables. Then we write

$$f_1(c_1,c_2) = \sum_{i=1}^{M_1} \{u(x_i) - w_i\}^2 \,, \tag{5.12}$$

where

$$u(b_1) = c_1 \tag{5.13}$$

$$v(b_1) = c_2 \tag{5.14}$$

$$\dot{u} = a_1 v \tag{5.15}$$

$$-\dot{v} = a_1 u . \tag{5.16}$$

The absorption coefficient a_1 is chosen so that

$$u(0) = 0 . \tag{5.17}$$

In addition, the principle of optimality yields the relationship

$$f_{K+1}(c_1, c_2) = \min_{a_{K+1}} \left\{ d_{K+1} + f_K(c_1', c_2') \right\} ,$$
$$K = 1, 2, \ldots, \tag{5.18}$$

$$d_{K+1} = \sum_i \{u(x_i) - w_i\}^2 , \tag{5.19}$$

with i ranging over integer values for which

$$b_K < x_i < b_{K+1} , \tag{5.20}$$

and

$$\dot{u} = a_{K+1} v, \; u(b_{K+1}) = c_1 \tag{5.21}$$

$$-\dot{v} = a_{K+1} u, \; v(b_{K+1}) = c_2 . \tag{5.22}$$

In addition we have put

$$c_1' = u(b_K) , \tag{5.23}$$

$$c_2' = v(b_K) . \tag{5.24}$$

In the usual manner of dynamic programming this leads to a computational scheme for computing the sequence of functions of two variables $f_1(c_1,c_2)$, $f_2(c_1,c_2)$, ..., and in principle solves our estimation problem. In the event that we do not wish to require that $u(0) = 0$, we may determine the function $f_1(c_1,c_2)$ this way:

$$f_1(c_1,c_2) = \min_{a_1} [\lambda u^2(0) + \sum_{i=1}^{M} \{u(x_i) - w_i\}^2] , \tag{5.25}$$

where λ is a suitably large parameter.

5-4 AN APPROXIMATE THEORY

While the original physical problem is a two-dimensional problem, it may be well-represented as a one-dimensional problem. Suppose that there are K segments of the medium and that the input is $v(b_K) = c$. The absorption coefficients $a_1, a_2, \ldots, a_K$ should be chosen to secure a minimum sum of squares of deviations from the measurements. Having picked the absorption coefficients, we may calculate the reflection coefficient $r(v(b_K))$ for this segmented medium. At the end b_K, the function u is essentially determined by v

and $r(v)$, $u(b_K) = v(b_K)\, r(v(b_K))$. The single variable $v(b_K) = c$ may then suffice to specify the state of the right end of the Kth segment.

Let us define the function $g_N(c)$

$$g_N(c) = \text{the smallest sum of squares of deviations on the first } N \text{ segments when the input is } c, \tag{5.26}$$

and the function $R_N(c)$,

$$R_N(c) = \text{the reflection coefficient that results when the optimal set of absorption coefficients is used on the first } N \text{ segments, the input being } c = v(b_N) . \tag{5.27}$$

The function $g_{N+1}(c)$ satisfies the inequality

$$g_{N+1}(c) \leq \min_{a} \{d_{N+1} + g_N(c')\} , \tag{5.28}$$

where

$$d_{N+1} = \sum_i [u(x_i) - w_i]^2, \; b_N < x_i < b_{N+1} , \tag{5.29}$$

and

$$\begin{aligned} \dot{u} &= a\, v, \quad v(b_{N+1}) = c\ , \\ -\dot{v} &= a\, u, \quad v(b_N) R_N(v(b_N)) = u(b_N)\ , \end{aligned} \tag{5.30}$$

and

$$c' = v(b_N)\ . \tag{5.31}$$

We do not have recurrence relations for the sequences of functions $g_N(c)$ and $R_N(c)$. We replace Eqs. (5.28), (5.29), and (5.30) by an approximate set, where instead of $g_N(c)$ we introduce the sequence $f_N(c)$, and instead of $R_N(c)$ we introduce $r_N(c)$. In our approximate theory we produce $f_{N+1}(c)$ from the recurrence formula

$$f_{N+1}(c) = \min_a \{d_{N+1} + f_N(c')\}\ , \tag{5.32}$$

where

$$d_{N+1} = \sum_i [u(x_i) - w_i]^2, \quad b_N < x_i < b_{N+1}\ , \tag{5.33}$$

and

$$c' = v(b_N)\ .$$

The following boundary value problem,

$$\begin{aligned} \dot{u} &= a\, v, \quad v(b_{N+1}) = c\ , \\ -\dot{v} &= a\, u, \quad v(b_N) r_N(v(b_N)) = u(b_N)\ , \end{aligned} \tag{5.34}$$

must be satisfied. The quantity

$$r_{N+1}(c) = r(b_{N+1}) \tag{5.35}$$

is obtained as the solution of the initial value problem

$$\dot{r} = a(1 + r^2), \; r(b_N) = r_N(c') \; . \tag{5.36}$$

For $N = 1$ we define

$$f_1(c) = \min_a \sum_i \alpha \, [u(x_i) - w_i]^2 \; , \tag{5.37}$$

where the summation is over indices for which

$$0 < x_i < b_1 \; , \tag{5.38}$$

and α is a weighting constant. Also we have

$$\begin{aligned} \dot{u} &= a\,v, \; u(0) = 0 \; , \\ -\dot{v} &= a\,u, \; v(b_1) = c \; , \end{aligned} \tag{5.39}$$

We define

$$r_1(c) = r(b_1) \tag{5.40}$$

where

$$\dot{r} = a(1 + r^2), \ 0 \le x \le b_1 , \tag{5.41}$$

$$r(0) = 0 .$$

The purpose of introducing the weight $\alpha \ge 1$ is to insure a good fit over the first segment.

Assuming that a unique minimizing solution exists, we can show that the results of our approximate theory are exact, if the observations w_i are perfectly accurate. We reason inductively. For the one segment process, there exists an input c_1^* for which $f_1(c_1^*) = 0$ by Eq. (5.37), and the reflection coefficient is $r_1(c_1^*)$. We assume that there exists an input to the medium of N segments, c_N^*, such that $f_N(c_N^*) = 0$, and that the reflection coefficient for this medium is $r_N(c_N^*)$. For the medium of $N+1$ segments, there is an input c_{N+1}^* such that $d_{N+1} = 0$, and the input (to the left) at b_N which satisfies condition (5.34) is $v(b_N) = c_N^*$. Therefore $f_{N+1}(c_{N+1}^*) = 0$, and the solution is exact.

In this manner we have reduced the original multidimensional optimization process to a sequence of one-dimensional processes.

5-5 A FURTHER REDUCTION

The solving of the nonlinear boundary value problem of Eq. (5.34) can be a source of difficulty. To aid in this process we note that we can write

$$v(b_N) = c\,T + u(b_N)\,R , \tag{5.42}$$

which follows simply from one of Chandrasekhar s invariance principles [1]. The transmission coefficient T and the reflection coefficient R of the $(N+1)^{st}$ segment are calculated from the solutions of the initial-value problems [6]

$$\dot{r} = a(1 + r^2), \quad r(0) = 0 \tag{5.43}$$

$$\dot{t} = a\, r\, t, \quad t(0) = 1 , \tag{5.44}$$

and

$$R = r(b_{N+1} - b_N) \tag{5.45}$$

$$T = t(b_{N+1} - b_N) . \tag{5.46}$$

In this way the second condition in Eq. (5.34) becomes

$$r_N(v)v = (v - c\, T)/R , \tag{5.47}$$

a nonlinear equation for $v = v(b_N)$.

5-6 COMPUTATIONAL PROCEDURE

The calculation of f_{N+1} for a given value of the parameter c may proceed as follows. We take a value of the coefficient a, and we produce numerically the reflection and transmission coefficients, R and T. Assuming we can solve Eq. (5.47) for $v(b_N)$, we go on to solve the linear two-point boundary-value problem of Section 5-4 by producing numerically two independent solutions of these homogeneous equations and determining constants so that the boundary conditions are met. Then the sum of squares of deviations d is computed, and the cost $\{d + f_N(v(b_N))\}$ is evaluated. We go through these steps for all the admissible choices of a, and the costs are compared. The value of a which makes the cost a minimum is the choice for the $(N+1)^{st}$ slab. The whole procedure is repeated for the range of values of c and of N.

It may be noted that in the calculation of the reflection coefficient r_{N+1}, the initial condition r_N is known only computationally on a grid of values of the argument. Experiments are needed to determine the required fineness of grid to achieve the required accuracy.

It is possible to derive recurrence relations for $f_N'(c)$ and $r_N'(c)$, and these can be employed in a variety of ways to improve the accuracy of the method. Numerical experimentation would have to be carried out to obtain reliable estimates of running times and accuracies. The method proposed here can be extended to treat the case where the interfact points are not known, though the computational effort will be greatly increased.

Experience with many similar problems leads us to believe that the proposed procedure is perfectly feasible [8, 9].

5-7 COMPUTATIONAL RESULTS

PRODUCTION OF OBSERVATIONS. We consider a homogeneous rod of unit length with absorption coefficient $a = 0.5$. We produce the internal fluxes to the right and to the left due to a unit input flux to the left at the end $x = 1$, and no input at the other end, $x = 0$. To do this, we use the fact that the quantity $v(1)$ is the reflection coefficient for the slab, which is $\tan a$ [6]. We integrate the transport equations with the initial values $u(1) = 1$, $v(1) = \tan a$, from $x = 1$ to $x = 0$. This procedure yields $u(x)$ and $v(x)$ throughout the rod.

TWO-DIMENSIONAL DYNAMIC PROGRAMMING PROCEDURE FOR THE DETERMINATION OF THE ABSORPTION COEFFICIENTS. The rod is divided up into 10 homogeneous sections of equal length. From the set of exact measurements, $w_j \approx u(x_j)$, we wish to determine the set of optimizing parameters a_N in each section. The correct solution is $a_N = 0.5$ for $N = 1, 2, \ldots, 10$.

In stage one of the multi-stage decision process, the rod is considered to consist of one segment extending from $x = 0$ to $x = 0.1$. If $c_1 = u(0.1)$, $c_2 = v(0.1)$, we choose the coefficient which makes $u(0) = 0$,

$$a = \frac{1}{0.1} \arctan \frac{c_1}{c_2}, \tag{5.48}$$

regardless of the measurements in this segment. The minimum cost is $f_1(c_1,c_2) = \sum_i [u(x_i) - w_i]^2$, where $u(x_i) = \sin a x_i$, $0 < x_i < 0.1$. This calculation is carried out for each value of c_1 and c_2.

The computations for the other stages $N = 2, 3, 4, \ldots$, may be best indicated by the following outline:

TWO-DIMENSIONAL DYNAMIC PROGRAMMING CALCULATIONS

For each stage $N = 2, 3, 4, \ldots,$

1. Print N
2. For each c_1
    1. For each c_2
        1. For each a
            1. Integrate to produce

               $c_1' = u(b_{N-1}), \; c_2' = v(b_{N-1}),$

               $$\begin{cases} \dot{u} = a\,v, & u(b_N) = c_1 \\ -\dot{v} = a\,u, & v(b_N) = c_2 \end{cases}$$

            2. Compute $d = \sum_i [u(x_i) - w_i]^2$
            3. Find $f_{N-1}(c_1', c_2')$ by interpolation
            4. Set $S(a) = d + f_{N-1}(c_1', c_2')$
        2. Search for $f_N(c_1, c_2) = \min \{S(a)\}$
        3. Print $c_1, c_2, a_N, c_1', c_2', f_N(c_1, c_2)$
3. For each c_1
    1. For each c_2
        1. Shift $f_N(c_1, c_2) \longrightarrow f_{N-1}(c_1, c_2)$.

There are four levels of computation: the stage N, the state c_1, the state c_2, the parameter a. The large brackets cover the steps which must be carried out at each level. By the statement "shift $f_N \to f_{N-1}$", we represent the discarding of the costs for stage $N-1$, and the replacement of f_{N-1} by the just computed costs for stage N, in readiness for the next stage. This saving in storage is allowed by the recurrence formula linking the current cost with the cost of only the previous stage. The interpolation may be carried out by the use of a linear formula in two dimensions, c_1 and c_2. The print-out value of a is, of course, the optimal value.

In our numerical trial, we execute the algorithm for three stages only, the rod then extending from $x = 0$ to $x = 0.3$. The exact observations are

$$
\begin{aligned}
u(0.02) &= 0.11394757 \times 10^{-1} \\
u(0.05) &= .28484388 \times 10^{-1} \\
u(0.08) &= .45567610 \times 10^{-1} \\
u(0.12) &= .68328626 \times 10^{-1} \\
u(0.15) &= .85381951 \times 10^{-1} \\
u(o.18) &= .10241607 \times 10^{0} \\
u(0.22) &= .12509171 \times 10^{0} \\
u(0.25) &= .14206610 \times 10^{0} \\
u(0.28) &= .15900853 \times 10^{0}
\end{aligned}
$$

The range of N is 1 to 3, the section interfaces lying at $x = 0.1, 0.2, 0.3$. The range of c_1 is 0.00(0.01) 0.20, 21 values; the range of c_2 is 1.120 (0.002) 1.140, 11 values. The five allowed values of a are 0.3 (0.1) 0.7. From the direct calculation, i.e., when the true structure of the rod is given, we know the conditions at the right end $x = 0.3$ which are $u(0.3) = 0.17028385$, $v(0.3) = 1.1266986$. The inverse calculations do not produce clearly the correct results $a_1 = a_2 = a_3 = 0.5$. It is believed that the grids of values of c_1 and of c_2 are not sufficiently fine, and that substantially improved results cannot be obtained without a great increase in computing expense. The computing time for the IBM 7044 is 1-1/2 minutes for these three stages. The one-dimensional reduction appears attractive in view of these results.

ONE-DIMENSIONAL DYNAMIC PROGRAMMING APPROXIMATION FOR THE DETERMINATION OF THE ABSORPTION COEFFICIENT. The rod of unit length is divided into five sections of equal length 0.2. Armed with the internal measurements $w_i \approx u(x_i)$, we wish to determine the absorption coefficient of each slab. The correct choices are $a_N = 0.5$ for $N = 1, 2, \ldots, 5$. In the one-dimensional case, the only state variable is $c = v(b_N)$.

The outline immediately following lists the calculations for producing a_1, $f_1(c)$ and $r_1(c)$ for $N = 1$, and the next outline shows the general scheme, $N = 2, 3, \ldots,$.

ONE-DIMENSIONAL DYNAMIC PROGRAMMING FOR STAGE N = 1

1. Print N
2. For each $c = v(b_N)$
   1. For each a
      1. Solve 2 point boundary-value problem for $v(0) = c'$,

         $$\begin{cases} \dot{u} = a\,v, \; u(0) = 0 \\ -\dot{v} = a\,u, \; v(0.2) = c \end{cases}$$

      2. Integrate to produce $u(x)$,

         $$\begin{cases} \dot{u} = a\,v, \; u(0) = 0 \\ -\dot{v} = a\,u, \; v(0) = c' \end{cases}$$

         and simultaneously calculate

         $d = \sum_i [u(x_i = w_i]^2$, and keep running estimate of $f_i(c) \simeq \min_a \{d\}$.
   2. Integrate to produce $r_1(c) = \rho(0.2)$,

      $$\dot{\rho} = a(1 + \rho^2), \; \rho(0) = 0$$

   3. Print $c, a_1, c', r_1(c), f_1(c)$.

ONE-DIMENSIONAL DYNAMIC PROGRAMMING FOR $N = 2, 3, 4, \ldots$

1. Print N
2. For each c
    1. For each a
        1. Produce $R = \rho(0.2)$, $T = t(0.2)$ by integrating

           $\dot{\rho} = a(1 + \rho^2)$, $\rho(0) = 0$

           $\dot{t} = a\,\rho\,t$, $t(0) = 1$

        2. Find $c' = v(b_{N-1})$ and $r_{N+1}(c')$ from N.L. eq.

           $R\,c' r_{N-1}(c') = c' - c\,T$

        3. Solve 2 point B.V.P. for $e' = u(0) = u(b_{N-1})$

           $\dot{u} = a\,v$, $v(0) = c'$

           $-\dot{v} = a\,u$, $v(0.2) = c$

        4. Integrate for $u(x)$:

           and calculate $d = \sum_i [u(x_i) = w_i]^2$

        5. Find $f_{N-1}(c')$ by interpolation
        6. Set $S(a) = d + f_{N-1}(c')$ and keep a running estimate of $f_N(c) \simeq \min_a \{S(a)\}$
    2. Integrate to produce $r_N(c) = \rho(0.2)$

       $\dot{\rho} = a(1 + \rho^2)$, $\rho(0) = r_{N-1}(c')$

    3. Print c, a_N, c', $r_N(c)$, $f_N(c)$
3. For each c
    1. Shift $f_N(c) \rightarrow f_{N-1}(c)$

To solve the nonlinear equation for $c' = v(b_{N-1})$ where $r(c')$ is known only on a grid of points, we compute the expressions $g_1 = R\,c'\,r_{N-1}(c')$, $g_2 = c' - c\,T$, and we take their difference $D = g_1 - g_2$. If $D = 0$, then c' has been found. Otherwise, we repeat the procedure for each discrete value c_i', until the sign of D_i is opposite to that of D_{i-1}. We then interpolate linearly to find the quantity c' which makes $D = 0$. If the sign of D does not change, i.e., the curves g_1 and g_2 do not intersect, then the corresponding value of a is definitely not allowed to be the coefficient for the segment in question.

If the minimum cost $f_N(c)$ is large for a given state and all remaining states may be deleted from further consideration. This provides a saving in computing time, for each state to be considered requires many calculations. Of course, the precaution must be taken to order the c's properly, so that no potentially vital state is lost.

The proposed one-dimensional scheme has been tested numerically. The range of N is 1 to 3, the interfaces of the sections being located at $x = 0.2$, 0.4, and 0.6. The states $v(b_N) = c$ are 1.04955 (0.00015) 1.13385, or 563 in all. This number is reduced in stage 2 to 546, by the use of the above test. Four values of the absorption coefficient a are allowed: 0.1, 0.3, 0.5, and 0.7. There are nine perfectly accurate observations of u per segment, a total of 27 data points. The integration method is Adams-Moulton with a grid size of 0.01.

From the output of our computation, we see that the minimum value of $f_3(c)$ is 0.387×10^{-8} and occurs when the input flux is $v(0.6) = c = 1.08855$ and the absorption coefficient for the segment of the medium between $x = 0.4$ and $x = 0.6$ is taken to be $a = 0.5$. This is very close to the true answer, $v(0.6) = 1.08860$, and the value of the parameter a is correct. The calculation tells us that the next state at $x = 0.4$ will be $v = 1.11673$. The nearest grid point in c is 1.11675, and the cost $f_2(1.11675)$ is indeed a minimum, 0.824×10^{-9}. The absorption coefficient for segment 2 is 0.5, the correct solution. The next state at $x = 0.2$ is predicted to be 1.13377. The nearest discrete state is 1.13385, possessing a cost $f_1(1.13385) = 0.181 \times 10^{-9}$. The absorption coefficient is 0.5, again the correct answer. The solutions at each state are clearly found, the minimum cost differing from the others by at least an order of magnitude. These dynamic programming calculations of about 20 minutes have very accurately determined the input, and they have identified the medium.

Now we wish to test the one-dimensional method of determining the structure of the medium when the measurements are few and of limited accuracy. We consider the rod of length 0.8 consisting of 4 segments of equal length 0.2. There are again the same 563 discrete states in c, and the same four possible absorption coefficients 0.1, 0.3, 0.5 and 0.7. However, there are only three observations per segment and these are correct to only two significant figures. Knowing the inputs to the first three stages $N = 1, 2, 3$, we see from the output of the calculations that the absorption coefficients are $a_1 = a_2 = a_3 = 0.5$, the correct solution in this region. On the other hand, we are not able to accurately identify the input to a given segment on the basis of these calculations, because the minimum of the function f is broad and it is not centered at the correct value of the input c. For stage $N = 4$, the value of a_4 is determined to be 0.3, and incorrect value. These experiments might serve as a warning to the experimental investigator. They show that the processing of data with a small number of measurements requires higher accuracy than two figures, and that if the measurements are of limited accuracy, many measurements should be made. This trial consumes 34 minutes of IBM 7044 computing time. This time of calculation could be reduced greatly by streamlining the calculations. No attempt to do this was made here; feasibility was our sole concern.

REFERENCES

1. Chandrasekhar, S., Radiative Transfer, Dover Publications, Inc., New York, 1960.

2. Agranovich, Z. S., and V. A. Marchenko, The Inverse Problem of Scattering Theory, Gordon and Breach, New York, 1963.

3. Bellman, R. E., H. Kagiwada, R. E. Kalaba, and Sueo Ueno, "On the Identification of Systems and the Unscrambling of Data - II: An Inverse Problem in Radiative Transfer," Proc. Nat. Acad. Sci., Vol. 53, pp. 910-913, May 1965.

4. Sims, A. R., "Certain Aspects of the Inverse Scattering Problem," J. Soc. Indust. Appl. Math., Vol. 5, pp. 183-205, 1957.

5. Bellman, R. E., Dynamic Programming, Princeton University Press, Princeton, New Jersey, 1957.

6. Bellman, R. E., and R. E. Kalaba, "Transport Theory and Invariant Imbedding," Nuclear Reactor Theory, pp. 206-208, edited by G. Birkhoff and E. Wigner, Amer. Math. Soc., Providence, Rhode Island, 1961.

7. Bellman, R. E., and R. E. Kalaba, "Functional Equations, Wave Propagatqon and Invariant Imbedding," J. of Math. and Mech., Vol. 8, September 1959, pp. 683-704.

8. Bellman, R. E., and S. E. Dreyfus, Applied Dynamic Programming, Princeton University Press, Princeton, New Jersey, 1962.

CHAPTER 6

SYSTEM IDENTIFICATION BY MEASUREMENT OF TRANSIENT WAVES

6-1 INTRODUCTION

The wave equation

$$\Delta u = \frac{1}{c^2} u_{tt}, \tag{6.1}$$

is one of the basic equations of mathematical physics. If we suppose that the local speed of propagation is a function of position

$$c = c(x, y, z) \tag{6.2}$$

then the difficulties in studying the various initial and boundary value problems which arise are well known [1, 2, 3]. In the sections which follow, we wish to study some of the inverse problems which arise when we attempt to determine the properties of a medium on the basis of observations of a wave passing through the medium. Such problems are of central importance in such varied areas as ionospheric and tropospheric physics, seismolology, and electronics. Some early results are due to Ambarzumian [4] and Borg [5].

We shall discuss some one-dimensional problems. Our basic technique is to reduce the partial differential equation in (6.1) to a system of ordinary differential equations either by using Laplace transforms or by considering the steady-state situation. Then our previously developed methodology is applicable. For simplicity and specificity we shall employ the nomenclature associated with the problem of the vibrating string. In passing, we note that our methodology is applicable to the diffusion equation, to the telegrapher's equation, and to other similar propagation equations.

6-2 THE WAVE EQUATION

Consider an inhomogeneous medium which extends from $x = 0$ to $x = 1$, for which the wave equation

$$u_{tt} = c^2 u_{xx} \tag{6.3}$$

is applicable. In this equation, the disturbance $u(x,t)$ is a function of position and time. Let us assume that the wave speed c satisfies the equation

$$c^2 = a + bx , \tag{6.4}$$

where a and b are constants, as yet unknown, which are to be determined on the basis of experiments.

Let the initial conditions be

$$u(x, 0) = g(x) , \tag{6.5}$$

$$u_t(x, 0) = v(x) . \tag{6.6}$$

Let the boundary conditions be

$$u(0, t) = 0 , \tag{6.7}$$

$$Tu_x(1, t) = f(t) . \tag{6.8}$$

Eqs. (6.3) - (6.8) may, for example, describe an inhomogeneous string, which is fixed at the end $x = 0$, while a force $f(t)$ is applied perpendicular to the string at $x = 1$, and T is the known tension.

The disturbance at the end $x = 1$, $u(1, t_i)$, is measured at N instants of time. On the basis of these observations, we wish to estimate the values of the parameters a and b, and thus to deduce the inhomogeneity of the medium.

6-3 LAPLACE TRANSFORMS

In order to reduce the partial differential wave equation to a system of ordinary differential equations, we take Laplace transforms of both sides of (6.3). We denote transforms by capital letters, for example,

$$U(x,s) = L\{u(x,t)\} = \int_0^\infty u(x,t)e^{-st}\, dt . \tag{6.9}$$

Equation (6.3) becomes

$$s^2U(x,s) - su(x,0) - u_t(x,0) = c^2U_{xx} . \tag{6.10}$$

Using (6.4) - (6.6), we obtain the desired system of ordinary differential equations,

$$(a + bx)U_{xx} = s^2U(x,s) - sg(x) - v(x) , \tag{6.11}$$

in which s is a parameter, $s = 1, 2., \ldots, N$. The boundary conditions are, from (6.7) and (6.8) ,

$$U(0,s) = 0, \quad TU_x(1,s) = F(s) . \tag{6.12}$$

The unknown constants a and b are to be determined by minimizing the expression

$$\sum_{s=1}^{N} [U_{Obs}(1,s) - U(1,s)]^2 . \tag{6.13}$$

The quantities $U_{Obs}(1,s)$ are the Laplace transforms of the experimentally observed values $u(1,t_i)$, while the quantities $U(1,s)$ are the solutions of equations (6.11) and (6.12). The use of Gaussian quadrature [6] leads to the approximate formula for the Laplace transform of the observations,

$$U_{Obs}(1,s) \cong \sum_{i=1}^{N} r_i^{s-1} u(1,t_i)w_i, \quad s = 1, 2, \ldots, N. \tag{6.14}$$

Similarly, the transform of the force may be produced with the use of the formula

$$F(s) \cong \sum_{i=1}^{N} r_i^{s-1} f(t_i) w_i, \quad s = 1, 2, \ldots, N. \tag{6.15}$$

In these equations, r_i are the roots of the shifted Legendre polynomial $P_N^*(x) = P_N(1 - 2x)$ and w_i are the related weights. In addition, the times of evaluation are

$$t_i = -\log_e r_i, \quad i = 1, 2, \ldots, N. \tag{6.16}$$

Interpolation may be necessary in order to have the data for these special times. After the solution has been found for $U(x,s)$, the inverse transforms $u(x,t)$ may be obtained by a numerical inversion method [7].

6-4 FORMULATION

The constants a and b are to be thought of as functions of x which satisfy the differential equations $a_x = 0$, $b_x = 0$. The complete system of equations for this nonlinear boundary value problem is,

$$\begin{aligned} U_{xx} &= \frac{1}{a+bx}\,[s^2U(x,s) - sg(x) - v(x)], \quad s=1, 2, \ldots, N, \\ a_x &= 0\,, \\ b_x &= 0\,. \end{aligned} \tag{6.17}$$

This is equivalent to a system of $2N + 2$ first order equations, so there must be $2N + 2$ boundary conditions. These conditions are

$$U(0,s) = 0, \quad s = 1, 2, \ldots, N , \tag{6.18}$$

$$U_x(1,s) = \frac{F(s)}{T} , \quad s = 1, 2, \ldots, N , \tag{6.19}$$

$$\frac{\partial}{\partial a} \left\{ \sum_{s=1}^{N} [U_{Obs}(1,s) - U(1,s)]^2 \right\} = 0 , \tag{6.20}$$

$$\frac{\partial}{\partial b} \left\{ \sum_{s=1}^{N} [U_{Obs}(1,s) - U(1,s)]^2 \right\} = 0 . \tag{6.21}$$

6-5 SOLUTION VIA QUASILINEARIZATION

The nonlinear boundary value problem may be resolved using the technique of quasilinearization [8, 9, 10]. In each step of the successive approximation method, we must solve the linear differential equations

$$\frac{dU_s^n}{dx} = W_s^n ,$$

$$\frac{dW_s^n}{dx} = \frac{s^2 U_s^n}{a+bx} - \frac{a^n + b^n x}{(a+bx)^2} s^2 U_s^n + \frac{s^2 U_s^n}{a+bx} , \tag{6.22}$$

$$\frac{da^n}{dx} = 0 ,$$

$$\frac{db^n}{dx} = 0 ,$$

where the superscripts n indicate the solution in the nth approximation, while the un-superscripted variables belong to the $(n-1)^{st}$ approximation. The boundary conditions are

$$U_s^n(0) = 0 , \tag{6.23}$$

$$W_s^n(1) = \frac{F(s)}{T} , \tag{6.24}$$

$$\frac{\partial}{\partial a^n} \left\{ \sum_{s=1}^{N} [U_{Obs}(1,s) - U_s^n(1)]^2 \right\} = 0 . \tag{6.25}$$

$$\frac{\partial}{\partial b^n} \left\{ \sum_{s=1}^{N} [U_{Obs}(1,s) - U_s^n(1)]^2 \right\} = 0 . \tag{6.26}$$

We represent the solution in the nth approximation as a linear combination of a particular vector solution and $N + 2$ homogeneous vector solutions. If we let the column vector $X(x)$ represent the solution in the nth approximation, where the components of X are $(U_1^n, U_2^n, \ldots, U_N^n, W_1^n, W_2^n, \ldots, W_N^n, a^n, b^n)$, and if we let the column vectors $P(x)$, $H^1(x)$, $H^2(x)$, $\ldots$, $H^{N+2}(x)$ represent the particular and homogeneous solutions, then we may write

$$X(x) = P(x) + \sum_{i=1}^{N+2} H^i(x) y_i . \tag{6.27}$$

Since the system of differential equations is of order $2N + 2$, and since N initial conditions are prescribed, there are $N + 2$ missing initial conditions, represented by the $N + 2$ dimensional column vector Y,

$$Y = (W_1^n(0),\ W_2^n(0),\ \ldots,\ W_N^n(0),\ a^n(0),\ b^n(0))^T . \tag{6.28}$$

The particular and homogeneous solutions are computationally produced. In terms of these, the boundary conditions (6.23) - (6.26) require the solution of system of $N + 2$ linear algebraic equations,

$$A\, Y = B , \tag{6.29}$$

where the elements of matrix A are

$$\begin{aligned} A_{ij} &= H_{N+1}^j(1) , && i = 1,\ 2, \ldots, N , \\ &= \sum_{s=1}^{N} H_s^j(1)\, H_s^i\,(1) , && i = N + 1, N + 2, \end{aligned} \tag{6.30}$$

$$j = 1, 2, \ldots, N, N + 1, N + 2,$$

and where the components of vector B are

$$\begin{aligned} B_i &= \frac{F(i)}{T} - P_{N+i}(1) , \quad i = 1, 2, \ldots, N , \\ &= \sum_{s=1}^{N} [U_{Obs}(1,s) - P_s(1)]\, H_s^i(1), \ i = N + 1, N + 2 . \end{aligned} \tag{6.31}$$

The method is applied iteratively for a fixed number of stages, about five, or it may be terminated when the approximations converge or diverge. The displacement function $u(x,t)$ may be obtained from its transform by a numerical inversion method of Bellman, _et al._ [7].

6-6 EXAMPLE 1 - HOMOGENEOUS MEDIUM, STEP FUNCTION FORCE

In this and the following example, we consider a homogeneous medium and make use of the analytical solution. In Example 3, we consider the more general problem of an inhomogeneous medium characterized by two unknown constants.

Consider the case in which we have a constant speed c which is given by the equation

$$c^2 = a = 1 . \tag{6.32}$$

The value of T is unity, the input $f(t)$ is the Heaviside unit step function, $H(t)$, and the initial conditions are $g(x) = v(x) = 0$. The wave equation for the function $U(x,s)$ is

$$U_{xx} = \frac{s^2}{c^2} U(x,s) . \tag{6.33}$$

The solution which satisfies (6.33), as well as the boundary conditions $U(0,s) = 0$, $TU_x(1,s) = F(s)$ is

$$U(x,s) = \frac{c\ F(s) \sinh \frac{s}{c} x}{T\ s \cosh \frac{s}{c}} . \tag{6.34}$$

Noting that the Laplace transform of the force is

$$F(s) = L\left\{H(t)\right\} = \frac{1}{s} , \tag{6.35}$$

we may explicitly evaluate U at the boundary $x = 1$, and we obtain the values

$$U(1,s) = \frac{c}{T}\frac{1}{s^2}\tanh\frac{s}{c} = \frac{1}{s^2}\tanh s . \tag{6.36}$$

The inverse transform,

$$u(1,t) = L^{-1}\left\{\frac{c}{T}\frac{1}{s^2}\tanh\frac{s}{c}\right\} , \tag{6.37}$$

is shown in Fig. 6-1.

We decide to use a seven point quadrature, so that $N = 7$. Making use of the known solution, we "produce" the observations at the specified times t_i, which are listed in Table 6-1.

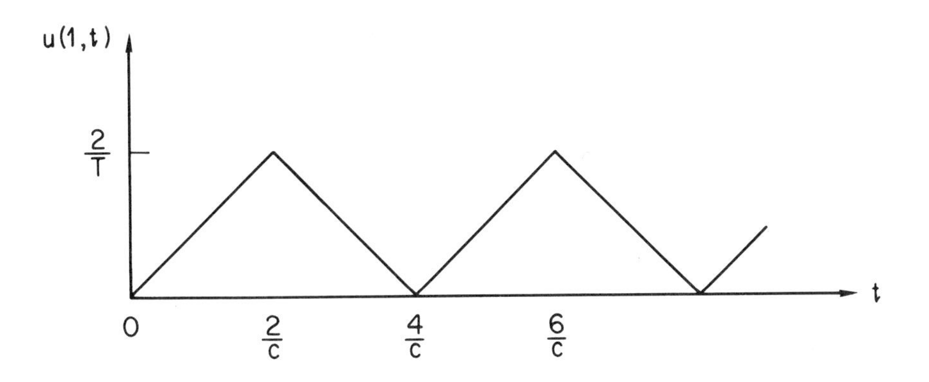

Figure 6-1 The Analytical Solution of the Wave Equation at $x = 1$, with a Step Function Input:
$u(1,t) = L^{-1} \{\frac{c}{T} \frac{1}{s^2} \tanh \frac{s}{c}\}$.

Table 6-1

Seven Observations for Example 1

t_i	$u(1,t_i)$
3.671195	0.328805
2.046127	1.953873
1.213762	1.213762
0.693147	0.693147
0.352509	0.352509
0.138382	0.138382
0.025775	0.025775

The approximate transforms U_{Obs} are computed using the formula from Gaussian quadrature. In Table 6-2, these quantities are compared with the exact transforms using the analytical solution. The transforms of the input, $F(s)$, are computed with the aid of the approximate formula. The approximate transforms, $U_{Obs}(1,s)$ and $F(s)$, are used in the calculations because in the general case, the analytical transforms will be unobtainable.

Table 6-2

THE LAPLACE TRANSFORMS $U_{Obs}(1,s)$ FOR EXAMPLE 1

s	Approximate $U_{Obs}(1,s)$	Exact $U_{Obs}(1,s)$
1	0.759442	0.761594
2	0.242907	0.241007
3	0.110753	0.110561
4	0.0624686	0.064580
5	0.0399963	0.0399964
6	0.0277773	0.0277774
7	0.0204081	0.0204081

There are only $2N + 1$ variables in this example, so that when $N = 7$, we have a solution of dimension 15. During each stage of the calculations, we have to produce a particular solution and $N + 1 = 8$ homogeneous solutions, i.e., $15x9 = 135$ differential equations must be integrated. For the initial conditions on P, we choose P(0) identically zero. We also choose for $H^j(0)$, the unit vector which has all of its components zero except the $(N+j)^{th}$, which is unity. Any linear combination of these P and H vectors identically satisfies the conditions $U_s(0) = 0$, $s = 1, 2, \ldots, N$. For the remaining boundary conditions, we must invert the 8x8 matrix A.

As a first check case, we try an initial approximation $a^0 = 1$ which is the correct value of a. We estimate the initial slopes to be $W_s(0) = 10^{-3}$. The initial approximation is generated by integrating the nonlinear equations with this set of estimates, as initial conditions. In three iterations we obtain better estimates of the slopes $W_s(0)$, but the value of a has drifted to 1.00023. This value may be used as a comparison for other trials. The results of three experiments are shown in the following table. The initial approximation a^0 is listed in Table 6-3, followed by the successive approximations a^n, $n = 1, 2, \ldots$, for each of the three trials.

Table 6-3

Successive Approximations of the Velocity a in Example 1

Approximation	Run 1	Run 2	Run 3
0	1.2	1.5	0.5
1	1.00991	0.46752	0.50922
2	1.00186	0.48284	0.80612
3	1.00018	0.67047	0.97736
4	-------	0.89366	1.00049
5	-------	0.99110	1.00022
6	-------	1.00041	-------

In Run 2, U and U_x at $x = 1.0$ are consistent to two significant figures with the conditions. In Run 3 U is in agreement with the observations to four places, and U_x agrees with the conditions to five figures. Recall that the conditions on U_x are supposed to be exact, and those on the $U(1,s)$ are of a least squares nature, which may help to explain why U_x is in better agreement than U for Run 3.

6-7 EXAMPLE 2 - HOMOGENEOUS MEDIUM, DELTA-FUNCTION FORCE

In Example 2, we have a homogeneous medium and zero initial conditions. The boundary conditions are again $u(0,t) = 0$, $u_x(1,t) = f(t)$, where now the input is $f(t) = \delta(t)$, the delta function. The Laplace transform of the delta function is $F(s) = 1$. The analytical solution for $x = 1$ is

$$u(1,t) = L^{-1}\left\{\frac{c}{T}\frac{1}{s}\tanh\frac{c}{s}\right\} . \tag{6.38}$$

This function is sketched in Fig. 6-2, for the case $c = 1$, $T = 1$.

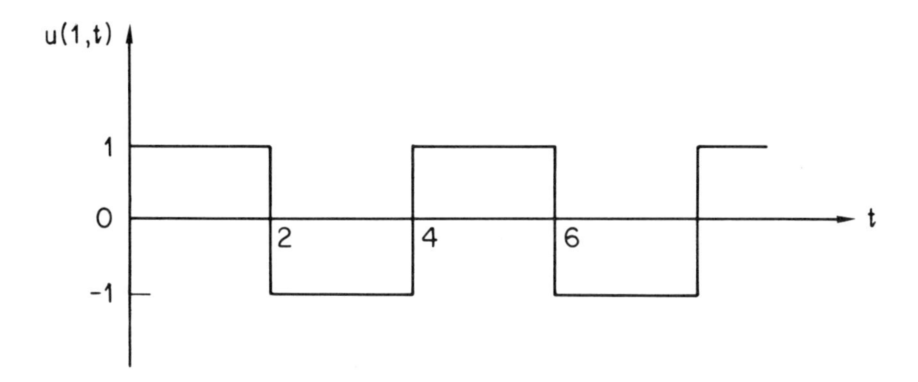

Figure 6-2 The Analytical Solution of the Wave Equation at $x = 1$, with a Delta Function Input: $u(1,t) = L^{-1}\{\frac{1}{s}\tanh s\}$.

We again take $N = 7$. The observations are

$$\begin{aligned} u(1,t_i) &= 1, \quad \text{for } i = 1, 2, \ldots, 5, \\ &= 1, \quad \text{for } i = 6, 7 . \end{aligned} \tag{6.39}$$

The transforms of the observations, $U_{Obs}(1,s)$, are computed using the quadrature approximation. A comparison of these values against the exact transforms using (6.38) is given in Table 6-4.

Table 6-4

THE LAPLACE TRANSFORMS $U_{Obs}(1,s)$ for EXAMPLE 2

s	Approximate $U_{Obs}(1,s)$	Exact $U_{Obs}(1,s)$
1	0.59080964	0.76159415
2	0.46055756	0.48201379
3	0.32857798	0.33168492
4	0.24939415	0.24983232
5	0.19992192	0.19998184
6	0.16665658	0.16666462
7	0.14285584	0.14285690

All initial approximations in the following experiments produced by integrating the nonlinear equations with a complete set of estimated initial conditions. The check case with initial approximation $a^0 = 1$, a correct guess, results in a convergence to the wrong value $a \cong 0.9$. With $a^0 = 0.5$, the estimate is again 0.9. With $a^0 = 1.5$, the value -0.8 is obtained. It is suspected that the discontinuous nature of the function $u(1,t)$ is the cause of the difficulty in determining a. A more reasonable formulation of the problem should include damping terms to overcome this obstacle. In spite of the poor estimates of a in the first two trials, the final approximations are quite close to the exact observations $U_{Obs}(1,s)$, rather than the approximate, and the conditions $U_x(1,s) = F(s)/T$ are met, to within 0.001%.

6-8 EXAMPLE 3 - INHOMOGENEOUS MEDIUM WITH DELTA-FUNCTION INPUT

As an example of the inverse problem for an inhomogeneous medium as originally posed, consider the case in which the wave velocity is indeed given by the equation

$$c^2 = a + bx , \tag{6.40}$$

where $a = 1$ and $b = 0.5$. We again set the initial conditions $u(x,0) = u_t(x,0) = 0$, and the tension $T = 1$. We exert a delta-function force, $f(t) = \delta(t)$, on the boundary $x = 1$, and we observe the displacement $u(1,t)$ as a function of time. Laplace transforms $U_{Obs}(1,s)$ are computed. The parameters a and b are determined for best agreement with these transforms of observations.

In this study, the experimenter obtains his data with the use of the digital computer, rather than by the actual performance of laboratory experiments. The exact solution for this inhomogeneous wave problem is not readily available analytically. We must produce the solution computationally, by solving the wave equation with its boundary conditions. Since we prefer to deal with the ordinary differential equation for the function $U_s(x)$, we solve the approximately equivalent linear two-point boundary value problem

$$U_{xx} = \frac{1}{a+bx} s^2 U(x,s) , \tag{6.41}$$

$$U(0, s) = 0 , \tag{6.42}$$

$$U_x(1,s) = 1 , \tag{6.43}$$

for $s = 1, 2, \ldots, N$. We produce a particular solution and N independent homogeneous solutions which, when combined to satisfy conditions (6.42) and (6.43), also produce the data of Table 6-5. These are the "observations".

Table 6-5

THE LAPLACE TRANSFORMS $U_{Obs}(1,s)$ FOR EXAMPLE 3

s	$U_{Obs}(1,s)$
1	.811967
2	.551174
3	.390695
4	.297835
5	.239837
6	.200613
7	.172392

These quantities $U_{Obs}(1,s)$ can be inverted numerically to produce the function $u(1,t_i) = L^{-1}\{U_{Obs}(1,s)\}$, which are the observations of the disturbance in the space of x and t. However, we need the set of transforms for use in determining the parameters a and b, and so we decide to utilize these numbers directly, as they appear in the table.

Two series of experiments are performed (see Tables 6-6 and 6-7). In one, the observations are given correct to 6 significant figures, and the initial approximations are varied. The true values of the unknown parameters are $a = 1.0$ and $b = 0.5$.

Table 6-6

SERIES I RESULTS FOR EXAMPLE 3

Observations are correct to six significant figures.

Run			
Run 1	$a^0 = .9$	$b^0 = .6$	$W_s^0(0)$ correct to 1 figure
	$a^3 = .9998$	$b^3 = .5002$	
Run 2	$a^0 = 1.2$	$b^0 = .3$	$W_s^0(0)$ correct to 1 figure
	$a^3 = .9998$	$b^3 = .5002$	
Run 3	$a^0 = 1.2$	$b^0 = .3$	$W_s^0(0) = .05$
	$a^5 = .9996$	$b^5 = .50005$	

Table 6-7

SERIES II RESULTS FOR EXAMPLE 3

Observations are in error by specified amounts

Run			
Run 4	$a^0 = 1.2$	$b^0 = .3$	$W_s^0(0)$ correct to 1 figure
	$a^3 = .9872$	$b^3 = .5182$	Observations, $\pm 1\%$ error
Run 5	$a^0 = 1.2$	$b^0 = .3$	$W_s^0(0)$ correct to 1 figure
	$a^3 = .937$	$b^3 = .590$	Observations: $\pm 5\%$ error

In the series I experiments, with accurate observations, rapid convergence to the correct values of the parameters occurs. The higher approximations of the initial slopes $W_s(0)$ are not listed, but these are considerably improved values.

In the series II experiments, noisy observations are used. For example in Run 4, the observations $U_{Obs}(1,s)$ are in error by the relative amounts +1%, -1%, +1%, ..., +1% for $s = 1, 2, 3, \ldots, 7$ respectively. The relative errors in the third approximations $a^3 = .9872$, $b^3 = .5182$, are 1.3% and 3.6% respectively. The results of this trial may be contrasted with the final approxmiations of Run 2. In Run 2, observations which are correct to six significant figures produce values of the parameters which are correct to less than 0.04%. Run 4 may also be compared with Run 6, in which case we are comparing the effect of 1% errors against 5% errors. The results of Run 6 involve errors of -6% in the value of a, and +18% in b.

The time required for these calculations is about one-half minute per iteration, with the IBM 7044. Each iteration includes the integration of $(N+3)(2N+2) = 10 \times 16 = 160$ differential equations, and the inversion of a 9x9 matrix.

6-9 DISCUSSION

The methods presented here are of practical use in identifying a system described by a wave equation or by linear differential equations or by a weighting function [11].

REFERENCES

1. Courant, R., and D. Hilbert, Methods of Mathematical Physics, Interscience Publishers, Inc., New York, Vol. 1, 1953, Vol. II, 1962.

2. Frank, P. and R. von Mises, Die Differential- und Integralgleichungen, Mary S. Rosenberg, N.Y., 1943.

3. Brillouin, L., Wave Propagation in Periodic Structures, Dover Publications, Inc., 1953.

4. Ambarzumian, V., "Über Einige Fragen der Eigenwerttheorie," Zeitschrift für Physik, Vol. 53, pp. 690-695, 1929.

5. Borg, G., "Eine Umkehrung der Sturm-Liouville Eigenwertaufgabe Bestimmung der Differentialgleichung Durch die Eigenwerte," Acta Math., Vol. 78, pp. 1-96, 1946.

6. Bellman, R., R. Kalaba, and M. Prestrud, Invariant Imbedding and Radiative Transfer in Slabs of Finite Thickness, American Elsevier Publishing Co., Inc., New York, 1963.

7. Bellman, R., H. Kagiwada, R. Kalaba, and M.C. Prestrud, Invariant Imbedding and Time-Dependent Transport Processes, American Elsevier Publishing Co., Inc., New York, 1964.

8. Kalaba, R., "On Nonlinear Differential Equations, the Maximum Operation, and Monotone Convergence," J. Math. and Mech., Vol. 8, pp. 519-574, 1959.

9. Bellman, R.E., and R. E. Kalaba, Quasilinearization and Boundary Value Problems, American Elsevier Publishing Co., New York, 1965.

10. Bellman, R., H. Kagiwada, and R. Kalaba, "Orbit Determination as a Multipoint Boundary-value Problem and Quasilinearization," Proc. Nat. Acad. Sci. USA, Vol. 48, pp. 1327-1329, 1962.

11. Bellman, R., H. Kagiwada and R. Kalaba, "Identification of Linear Systems via Numerical Inversion of Laplace Transforms", IEEE Transactions on Automatic Control, Vol. AC-10, No. 1, pp. 111-112, January 1965.

CHAPTER 7

STEADY STATE WAVE PROPAGATION

7-1 INTRODUCTION

Consider the propagation of waves in a plane parallel stratified medium [1-5] extending from $x = 0$ to $x = b$, with index of refraction $n(x)$ varying continuously throughout the slab. The slab is bounded by a vacuum to the left ($n_0 = 1$) and a homogeneous medium with index of refraction n_1 to the right, as shown in Fig. 7-1. We assume a lossless dielectric medium in which $n(x)$ is independent of frequency.

The wave equation is

$$k^2(x)u_{tt} = u_{xx} , \tag{7.1}$$

where k is the wave number. The wave number is related to the index of refraction by the formula

$$k(x) = \frac{\omega}{c_0} n(x) , \tag{7.2}$$

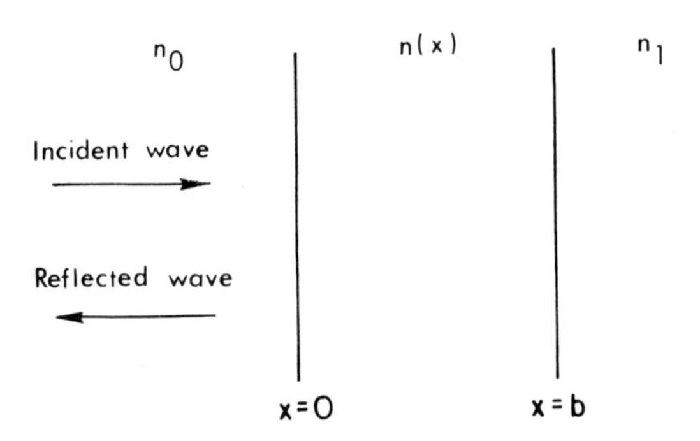

Figure 7-1 The Physical Situation

where c_0 is the speed of light in a vacuum, which we normalize to unity, and ω is the angular frequency. We are interested in solutions of the form

$$u(x,t) = e^{-i\omega t}u(x) , \tag{7.3}$$

corresponding to the steady-state case where the transients have died down. The function $u(x)$ satisfies the ordinary differential equation.

$$u''(x) + \omega^2 k^2(x)u(x) = 0 . \tag{7.4}$$

We shall often neglect the function $e^{-i\omega t}$ in all of the solutions, and speak of the functions $u(x)$ as waves.

We conduct a series of experiments in which waves of different frequencies ω_j are normally incident in the medium from the left, i.e., the incident wave is $e^{i(k_0 -x-\omega_j t)}$, or simply $e^{ik_0 x}$. The reflected waves at each frequency are observed. We wish to determine the index of refraction $n(x)$

throughout the slab on the basis of these measurements.

7-2 SOME FUNDAMENTAL EQUATIONS [1, 2]

Consider the case of two adjacent homogeneous media, as illustrated in Fig. 7-2.

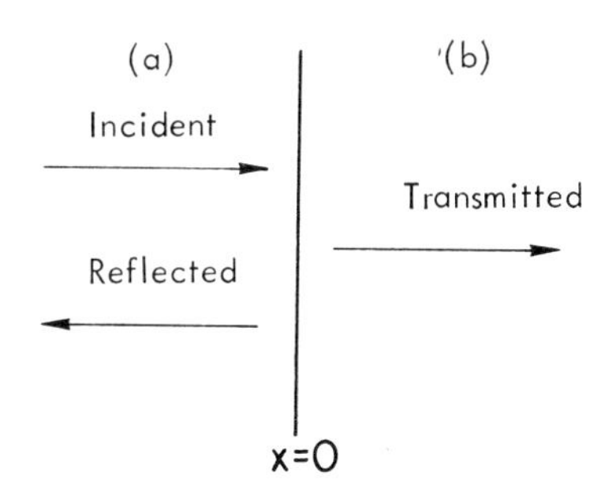

Figure 7-2 Waves At An Interface

A plane wave of frequency ω traveling in medium (a) is incident at the interface $x = 0$. Let the wave numbers of medium (a) and medium (b) be k_a and k_b respectively. The incident wave is $e^{ik_a x} e^{-i\omega t}$ and the reflected wave is $re^{ik_a x} e^{-i\omega t}$ where

$$r = \frac{k_a - k_b}{k_a + k_b} . \tag{7.5}$$

The transmitted wave is $te^{ik_b x} e^{-i\omega t}$, where

$$t = \frac{2k_a}{k_a + k_b} . \tag{7.6}$$

Now consider the case of two interfaces between three homogeneous media, (a), (b), and (c). The interfaces are separated by a distance Δ. See Fig. 7-3.

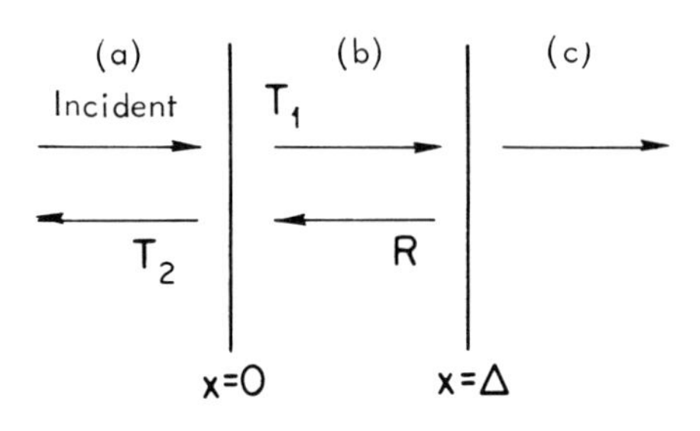

Figure 7-3 Waves At Two Interfaces

The incident wave is again $e^{ik_a x} e^{-i\omega t}$. The wave which is transmitted through $x = 0$, reflected at $x = \Delta$, and transmitted again through $x = 0$ is $v e^{-ik_a x} e^{-i\omega t}$, where

$$v = \frac{2k_a}{k_a + k_b} \cdot \frac{k_b - k_c}{k_b + k_c} e^{2ik_b\Delta} \cdot \frac{2k_b}{k_b + k_a} + o(\Delta) , \tag{7.7}$$

and $o(\Delta)$ includes the terms proportional to the second and higher powers of Δ. This equation shows how v depends on frequency by means of the exponential factor
$e^{2ik_b\Delta} = e^{2i\omega n_b\Delta}$.

7-3 INVARIANT IMBEDDING AND THE REFLECTION COEFFICIENT [5]

Now we turn our attention to the reflection coefficient r as a function of thickness of the medium. We assume that the slab is inhomogeneous and that it extends from $x = z$ to $x = b$. The right boundary $x = b$ is to be considered fixed, while the left boundary $x = z$ is variable, as shown in Fig. 7-4. The incident wave is $e^{ik(z\_)(x-z)}$, deleting the time dependent factor $e^{-i\omega t}$, where $k(z\_) = k(z-0)$ is the wave number of the homogeneous medium to the left, and where the expression $e^{ik(z\_)(x-z)}$ is used rather than $e^{ik(z\_)x}$ in order to normalize the incoming intensity at $x = z$.

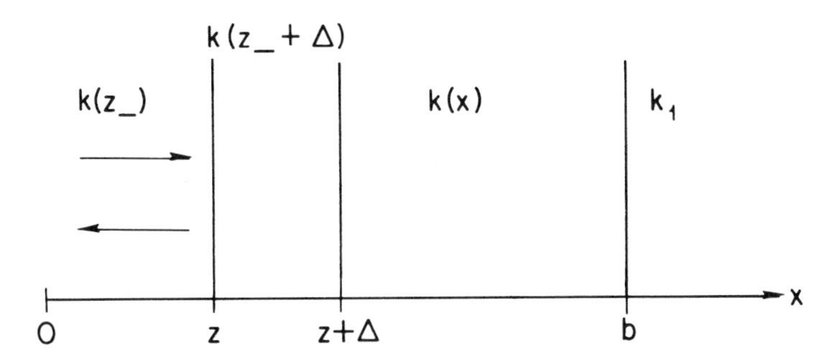

Figure 7-4 An Inhomogeneous Medium of Thickness $b-z$

The reflected wave is $r(z)e^{-ik(z\_)(x-z)}$.

Using the technique of invariant imbedding, we relate the reflection coefficient for a slab extending from z to b to that for a slab extending from $z + \Delta$ to b. The reflected wave may be expressed, to terms of order zero and one in Δ, as arising from three processes:

(a) immediate reflection at z;

(b) transmission through the interface at $x = z$, reflection at $z + \Delta$ from the slab $(z + \Delta, b)$, and transmission through z;

(c) transmission through the interface at $z = z$, reflection at $z + \Delta$ from the slab $(z + \Delta, b)$, reflection at z, reflection at $z + \Delta$, and finally transmission through z.

These three cases are represented in Fig. 7-5.

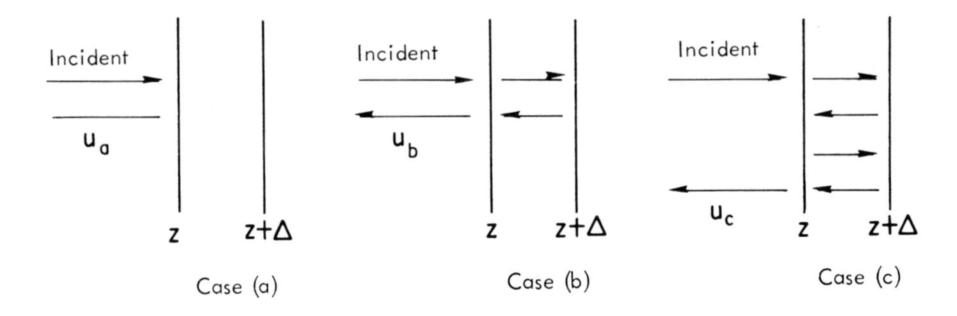

Figure 7-5 Three Processes in a Stratified Slab

The wave which is reflected from the slab (z,b) is

$$r(z)e^{-ik(z_-)(x-z)} = [r_a + r_b + r_c + o(\Delta)]e^{-ik(z_-)(x-z)} , \tag{7.8}$$

where

$$r_a = \frac{k(z_-) - k(z_-+\Delta)}{k(z_-) + k(z_-+\Delta)} , \tag{7.9}$$

$$r_b = \frac{2k(z_-)}{k_{z_-}) + k(z_-+\Delta)} \cdot r(z+\Delta)e^{2ik(z_-+\Delta)} \cdot \frac{2k(z_-+\Delta)}{k(z_-+\Delta) + k(z_-)} , \tag{7.10}$$

$$r_c = \frac{2k(z_-)}{k(z_-) + k(z_-+\Delta)} \cdot r(z+\Delta)e^{2ik(z_-+\Delta)\Delta}$$

$$\cdot \frac{k(z_-+\Delta) - k(z_-)}{k(z_-+\Delta) + k(z_-)} \cdot r(z+\Delta)e^{2ik(z_-+\Delta)\Delta} \tag{7.11}$$

$$\cdot \frac{2k(z_-+\Delta)}{k(z_-+\Delta) + k(z_-)}$$

and $k(z_-+\Delta) = k(z+\Delta-0)$ is the wave number in the region immediately to the left of the interface $z+\Delta$. Simplifying to terms of order Δ, we have

$$r(z) = \frac{k(z_-)-k(z_-+\Delta)}{[k(z-)+k(z_-+\Delta)]}$$

$$+ \frac{4k(z_-)k(z_-+\Delta)}{[k(z_-)+k(z_-+\Delta)]^2} [1+2ik(z_-+\Delta)\Delta]r(z+\Delta)$$

$$- \frac{4k(z_-)k(z_-+\Delta)}{[k(z_-)+k(z_-+\Delta)]^2}$$

$$\cdot \frac{[k(z_-)-k(z_-+\Delta)]}{[k(z_-)+k(z_-+\Delta)]} [1+4ik(z_-+\Delta)\Delta]r^2(z+\Delta)$$

$$+ o(\Delta) . \tag{7.12}$$

Making use of the formula for the derivative of r,

$$\frac{dr}{dz} = \lim_{\Delta\to 0} \frac{r(z+\Delta)-r(z)}{\Delta} , \tag{7.13}$$

we obtain the Riccati equation

$$\frac{dr}{dz} = \frac{k'}{2k} - 2ikr - \frac{k'}{2k} r^2 . \tag{7.14}$$

The "initial" condition reduces to the formula for an interface between two media

$$r(b) = \frac{k(b-0)-k_1}{k(b-0)+k_1} . \tag{7.15}$$

In terms of the index of refraction, Eqs. (7.14) and (7.15) are

$$\frac{dr}{dz} = \frac{n'}{2n} - 2in\omega r - \frac{n'}{2n} r^2 , \tag{7.16}$$

where $n = n(z)$, and

$$r(b) = \frac{n(b)-n_1}{n(b)+n_1} . \tag{7.17}$$

The reflection coefficient for any inhomogeneous slab in which n varies as a function of x may be found by a simple (numerical) integration of (7.16) with the given initial condition (7.17). The integration is carried out from the right boundary $z = b$ to the left boundary $z = 0$.

7-4 PRODUCTION OF OBSERVATIONS

In place of performing laboratory experiments for obtaining reflection data [6, 7], we produce the observations computationally, for N different frequencies. The incident waves are $e^{i\omega_j n_0 x}$, and the reflected waves are

$r_j(0)e^{-i\omega_j n_0 x}$, $j = 1, 2, \ldots, N$. We solve the initial value problems

$$\frac{dr_j}{dz} = \frac{n'}{2n} - 2in\omega_j r_j - \frac{n'}{2n} r_j^2 , \tag{7.18}$$

$$r_j(b) = \frac{n(b)-n_1}{n(b)+n_1} , \quad b \geq z \geq 0 , \tag{7.19}$$

for the desired coefficients $r_j(0)$.

Since r_j is a complex reflection coefficient, we let

$$r_j = R_j + iS_j , \tag{7.20}$$

where R_j and S_j are real functions which satisfy the differential equations

$$\frac{dr_j}{dz} = \frac{n'}{2n} + 2n\omega_j S_j - \frac{n'}{2n} (R_j^2 - S_j^2) ,$$

$$\frac{dS_j}{dz} = 2n\omega_j R_j - \frac{n'}{n} R_j S_j , \tag{7.21}$$

and

$$R_j(b) = \frac{n(b)-n_1}{n(b)+n_1} , \quad S_j(b) = 0 , \tag{7.22}$$

for $j = 1, 2, \ldots, N$.

For the numerical experiment, we take

$$n(x) = a_1 + a_2(x-1)^2 \tag{7.23}$$

where $a_1 = 1$, $a_2 = 0.5$. We also choose

$$\begin{aligned} b &= 1 \,, \\ N &= 3 \,, \\ \omega_1 &= 2\pi \,, \\ \omega_2 &= 4\pi \,, \\ \omega_3 &= 6\pi \,. \end{aligned} \tag{7.24}$$

We assume that $n_1 = n(b)$, so that $R_j(b) = 0$.

We have chosen to normalize the speed,

$$c_0 = 3 \times 10^{10} \text{cm/sec} = \frac{\text{one length unit}}{\text{one time unit}} \,. \tag{7.25}$$

We have chosen

$$b = 1 \text{ length unit} \tag{7.26}$$

and we set

$$\begin{aligned} b &= 3 \text{ cm} \\ &\cong 1 \text{ X-band microwave wave length.} \end{aligned} \tag{7.27}$$

Then

$$1 \text{ length unit} = 3 \text{ cm} \tag{7.28}$$

and

$$1 \text{ time unit} = 10^{-10} \text{ sec.} \tag{7.29}$$

To produce R_j and S_j, the real and imaginary parts of the reflection coefficients, for incident waves of frequencies 10, 20, and 30 gigahertz, we integrate Eqs. (7.21) with initial conditions $R_j(1) = 0$, $S_j(1) = 0$, for $j = 1, 2, 3$. We use a step length of $-.001$ and the Adams-Moulton integration scheme. The values $R_j(0)$, and $S_j(0)$ are the "observed" reflection coefficients. These are

$$\begin{aligned} R_1(0) &= .13217783 \times 10^{-2}, & S_1(0) &= .14843017 \times 10^{-1}, \\ R_2(0) &= .32313148 \times 10^{-3}, & S_2(0) &= .95414704 \times 10^{-2}, \\ R_3(0) &= .38854984 \times 10^{-3}, & S_3(0) &= .58976205 \times 10^{-2}. \end{aligned} \tag{7.30}$$

7-5 DETERMINATION OF REFRACTIVE INDEX

We consider the inhomogeneous slab extending from $x = 0$ to $x = 1$. We are given observations of the real and imaginary parts of the reflection coefficients, $A_i \cong R_i$, $B_i \cong S_i$, where

$$\begin{aligned} A_1 &= .132178 \times 10^{-2}, & B_1 &= .148430 \times 10^{-1}, \\ A_2 &= .323131 \times 10^{-3}, & B_2 &= .954147 \times 10^{-2}, \\ A_3 &= -.388550 \times 10^{3}, & B_3 &= .589762 \times 10^{-2}, \end{aligned} \tag{7.31}$$

which correspond to frequencies $\omega_1 = 10$, $\omega_2 = 20$, and $\omega_3 = 30$ gigahertz [6, 7]. We seek to determine the values of the constants a and b in the equation for the index of refraction as a function of position,

$$n(x) = a + b(x-1)^2 , \tag{7.32}$$

in such a manner as to minimize the expression

$$S = \sum_{i=1}^{3} [(A_i - R_i(0))^2 + (B_i - S_i(0))^2] . \tag{7.33}$$

The form S is the sum of squares of deviations between the solution of Eqs. (7.21) and (7.22), and the (perhaps inaccurate) observations (7.31).

The system of nonlinear equations is

$$\begin{aligned} R_j' &= \frac{n'}{2n} + 2n\omega_j S_j - \frac{n'}{2n}(R_j^2 - S_j^2) \\ S_j' &= -2n\omega_j R_j - \frac{n'}{n} R_j S_j, \quad j = 1, 2, 3, \\ a' &= 0 , \\ b' &= 0, \end{aligned} \tag{7.34}$$

where

$$n = a + b(x-1)^2 , \tag{7.35}$$

and

$$n' = 2b(x-1) . \tag{7.36}$$

We obtain a system of linear differential equations by applying quasilinearization. In the following linear equations, so as not to clutter the equations with superscripts indicating the approximations and subscripts indicating the components, we write the variables of the current k^{th} approximation as R, S, a, b, (also n and n'). Corresponding quantities in the previous $(k-1)^{st}$ approximation are ρ, σ, α, β (and η and η'). The linear equations obtained via quasilinearization are

$$R' = \frac{\eta'}{2\eta} + 2\eta\omega\sigma - \frac{\eta'}{2\eta}(\rho^2-\sigma^2) + (R-\rho)(-\frac{\eta'}{\eta}\rho)$$

$$+ (S-\sigma)\ (2\eta\omega + \frac{\eta'}{\eta}) + (a-\alpha)[\frac{1}{2}\frac{\partial}{\alpha\alpha}(\frac{\eta'}{\eta})$$

$$+ 2\omega\alpha\frac{\partial\eta}{\partial\alpha} - \frac{1}{2}(\rho^2-\sigma^2)\frac{\partial}{\partial\alpha}(\frac{\eta'}{\eta})] + (b-\beta)\ [\frac{1}{2}\frac{\partial}{\partial\beta}(\frac{\eta'}{\eta})$$

$$+ 2\omega\sigma\frac{\partial\eta}{\partial\beta} - \frac{1}{2}(\rho^2-\sigma^2)\frac{\partial}{\partial\beta}(\frac{\eta'}{\eta})]\ , \tag{7.37}$$

$$S' = -\ 2\eta\omega\rho - \frac{\eta'}{\eta}\rho\sigma + (R-\rho)(-2\eta\omega - \frac{\eta'}{\eta}\sigma)$$

$$+ (S-\sigma)(-\frac{\eta'}{\eta}\rho) + (a-\alpha)[-2\omega\rho\frac{\partial\eta}{\partial\alpha} - \rho\sigma\frac{\partial}{\partial\alpha}(\frac{\eta'}{\eta})]$$

$$+ (b-\beta)[-2\omega\rho\frac{\partial\eta}{\partial\beta} - \rho\sigma\frac{\partial}{\partial\beta}(\frac{\eta'}{\eta})] \tag{7.38}$$

$$a' = 0\ , \tag{7.39}$$

$$b' = 0\ . \tag{7.40}$$

In these equations, we must make the substitutions

$$\frac{\partial \eta}{\partial \alpha} = 1, \qquad \frac{\partial}{\partial \alpha}\left(\frac{\eta'}{\eta}\right) = -\frac{\eta'}{\eta^2},$$

$$\frac{\partial \eta}{\partial \beta} = (x-1)^2, \quad \frac{\partial}{\partial \beta}\left(\frac{\eta'}{\eta}\right) = 2\frac{x-1}{\eta} - \frac{\eta'}{\eta^2}(x-1)^2 . \tag{7.41}$$

For each iteration of the successive approximation scheme, we produce numerically a particular vector solution $p(x)$ and two homogeneous vector solutions $h^1(x)$ and $h^2(x)$ of the system (7.38) - (7.39). We set the components of the reflection coefficients equal to a linear combination of the components of $p(x)$, $h^1(x)$, and $h^2(x)$,

$$R_j^k = p_j(x) + a\, h_j^1(x) + b\, h_j^2(x), \quad j = 1, 2, 3,$$

$$S_j^k = p_{j+3}(x) + a\, h_{j+3}^1(x) + b\, h_{j+3}^2(x), \quad j = 1, 2, 3,$$

$$a^k = p_7(x) + a\, h_7^1(x) + b\, h_7^2(x) = a,$$

$$b^k = p_8(x) + a\, h_8^1(x) + b\, h_8^2(x) = b. \tag{7.42}$$

The multipliers a and b are given by the equations

$$\frac{\partial}{\partial a}\left\{\sum_{i=1}^{3} [(A_i - R_i^k(0))^2 + (B_i - S_i^k(0))^2)\right\} = 0,$$

$$\frac{\partial}{\partial b}\left\{\sum_{i-1}^{3} [(A_i - R_i^k(0))^2 + (B_i - S_i^k(0))^2]\right\} = 0. \tag{7.43}$$

After making the substitutions (7.42), we obtain the values of a and b in the current approximation,

$$a = (f_1 e_{22} - f_1 e_{22}) / (e_{11} e_{22} - e_{12} e_{21}) ,$$
$$b = (e_{11} f_1 - e_{21} f_1) / (e_{11} e_{22} - e_{12} e_{21}) , \tag{7.44}$$

where the right hand sides are given in terms of known quantities,

$$f_i = \sum_{\ell=1}^{3} h_\ell^i(0)(A_\ell - P_\ell(0)) + \sum_{\ell=1}^{3} h_{\ell+3}^i(0)(B_\ell - P_{\ell+3}(0))$$

$$e_{ij} = \sum_{\ell=1}^{6} h_\ell^i(0) h_\ell^j(0) , \quad j = 1, 2 ,$$

$$i = 1, 2 . \tag{7.45}$$

7-6 NUMERICAL EXPERIMENTS

Using the given data, and the initial approximation for refractive index $n(x) = 1.2 + 0.2(x-1)^2$, we determine the constants a and b in the function $n(x) = a + b(x-1)^2$ to one part in 10^6 after five iterations of quasilinearization. The successive approximations of the constants a and b are listed in Table 7-1, labelled Trial 1, and the approximations of the index of refraction are shown in Fig. 7-6.

For the next experiment, we use data which are in error by $\pm 2\%$:

$$\begin{aligned} A_1 &= .134822 \times 10^{-2}, & B_1 &= .145461 \times 10^{-1}, \\ A_2 &= .316668 \times 10^{-3}, & B_2 &= .935364 \times 10^{-2}, \\ A_3 &= -.396321 \times 10^{-3}, & B_3 &= .601557 \times 10^{-2}. \end{aligned} \tag{7.46}$$

Table 7-1

SUCCESSIVE APPROXIMATIONS OF THE PARAMETERS
a AND b IN THE EQUATION FOR REFRACTIVE INDEX

Approximation	Trial 1		Trial 2	
	a	b	a	b
0	1.2	0.2	1.2	0.2
1	0.9570026	0.3901124	0.9634062	0.3833728
2	1.0231246	0.4476300	1.0249423	0.4388065
3	1.0024801	0.4955928	1.0085360	0.4839149
4	0.9999989	0.4999477	1.0031811	0.4849384
5	0.9999993	0.4999996	1.0034835	0.4851507

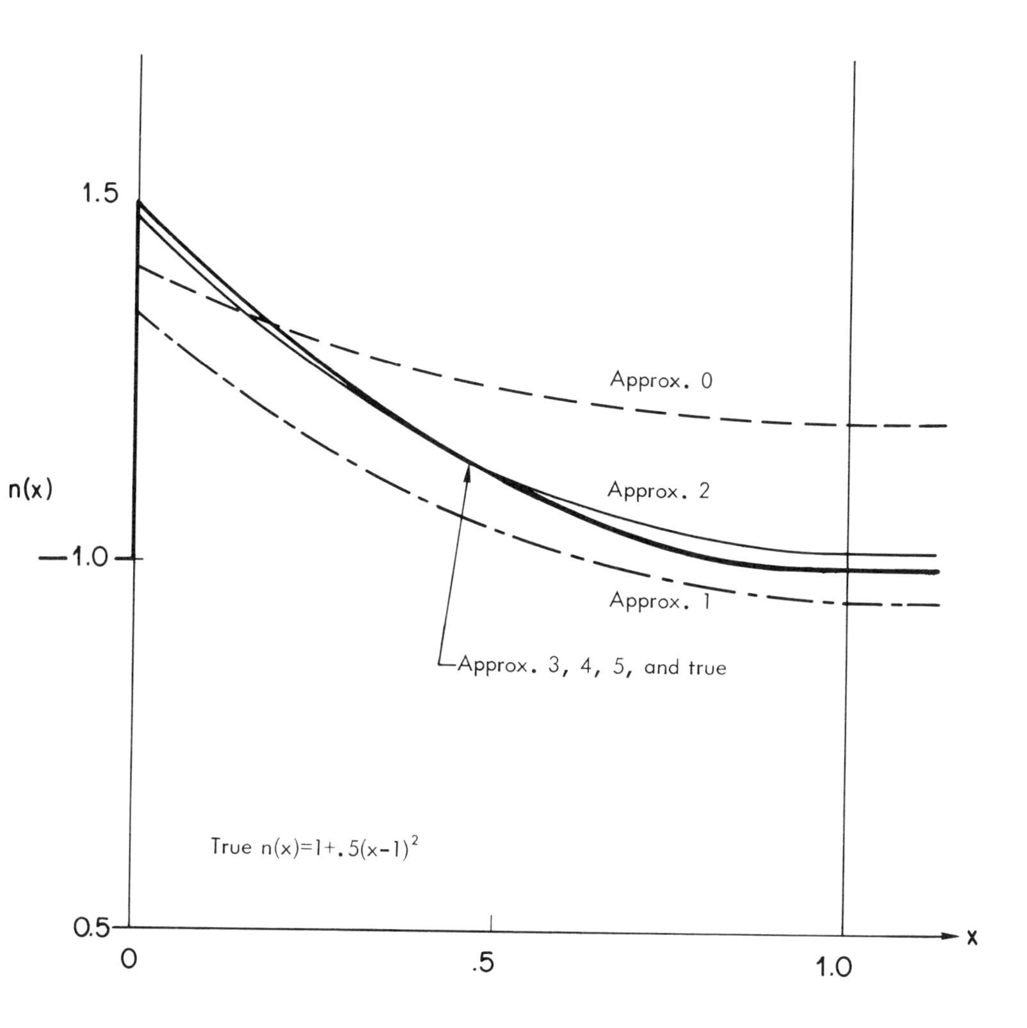

Figure 7-6 Successive Approximation of the Index of Refraction, Trial 1.

After five iterations, the initial approximation being the same as before, the constant a is found correct to within 0.3%, and b is correct to about 3%. On the other hand, the error in $n(x)$ ranges from 0.3% at $x = 1$ to only 0.7% at $x = 0$. The results are given in Table 7-1.

For each trial, the step length of integration is -.0025, and the integration scheme is Adams-Moulton. The time of calculations is 2 min. 12 sec. on the IBM 7044.

7-7 DISCUSSION

Inverse problems in wave propagation, as well as in particle processes, can be computationally solved. The wave equation, being a partial differential equation, is replaced by a system of ordinary differential equations in one of several ways. In the previous chapter, we used Laplace transform methods. In this chapter, we assumed a solution of the form $u(x,t) = u(x)\, e^{-i\omega t}$, and we obtained ordinary differential equations for $u(x)$. Another Fourier decomposition might be

$$u(x,t) = \sum_{n=1}^{N} a_n(x) \sin nt \,, \tag{7.47}$$

which results in second order ordinary differential equations for the functions $a_n(x)$. Another system of ordinary differential equations results when the space derivative is replaced by a finite difference,

$$\ddot{u}_n(t) \cong \frac{1}{c^2} \frac{u_{n+1}(t) - 2\, u_n(t) + u_{n-1}(t)}{\Delta^2} \tag{7.48}$$

These offer interesting possibilities for further studies.

REFERENCES

1. Lindsay, R.B., Mechanical Radiation, McGraw-Hill Book Company, Inc., New York, 1960.

2. Morse, P. M., Vibration and Sound, McGraw-Hill Book Company, Inc., New York, 1948.

3. Osterberg, H., "Propagation of Plane Electromagnetic Waves in Inhomogeneous Media," J. Opt. Soc. America Vol. 48, pp. 513-521, 1958.

4. Schelkunoff, S. A., "Remarks Concerning Wave Propagation in Stratified Media," The Theory of Electromagnetic Waves, A Symposium, Interscience Publishers, Inc., New York, pp. 181-192, 1951.

5. Bellman, R., and R. Kalaba, "Functional Equations, Wave Propagation and Invariant Imbedding," J. Math. Mech. Vol. 8, pp. 683-704, 1959.

6. Blore, W. E., P. E. Robillard, and R. I. Primich, "35 and 70 Gc Phase-Locked CW Balanced-Bridge Model Measurement Radars," Microwave Journal, Vol. 7, 61-65, 1964.

7. Kyhl, R. L., "Directional Couplers," Technique of Microwave Measurements, edited by C. G. Montgomery, McGraw-Hill Book Company, Inc., New York, pp. 854-897, 1947.

Chapter 8

PROBING AN INHOMOGENEOUS MEDIUM WITH RAYS

8-1 INTRODUCTION

In the fields of seismology, oceanology, meteorology, and physics in general, a basic problem is to determine the properties of an inhomogeneous medium based on observation and analysis of a propagation process [1-5]. In the ray approximation, this inverse problem can be directly formulated as a nonlinear boundary-value problem. Another advantage of the ray treatment over that of the wave equation is that a medium whose properties vary in two or three dimensions may be treated.

There are well-known difficulties associated with the solution of the original partial differential wave equation [6]. However, it is possible to approximate the wave equation by a system of ordinary differential equations by the use of Laplace transforms and by the method of lines [7], as well as by the use of rays. Integrodifferential equations can also be reduced to systems of ordinary differential equations by the use of finite ordinate methods, as shown in earlier chapters.

8-2 STATEMENT OF THE PROBLEM

Consider a propagation process in an infinite and inhomogeneous plane medium whose properties vary in the x and y directions. It is supposed that some knowledge about the structure of the medium exists so that the index of refraction may be expressed not only as a function of x and y, but also as dependent on two parameters, a and b [1, 2].

$$n = n(x, y, a, b) \tag{8.1}$$

for $0 \leq x, y \leq 1$, $a \cong \alpha$, and $b \cong \beta$, where α and β are initial estimates of the parameters. For simplicity the reference speed is taken as unity.

It is assumed that the ray approximation is appropriate. In Fig. 8-1, we sketch some of the N rays emanating from the origin and ending at various points on the line $x = 1$. Regard the origin as the location of a transmitter and the points $(1, Y^i)$ as the locations of receivers $(i = 1, 2, \ldots, N)$.

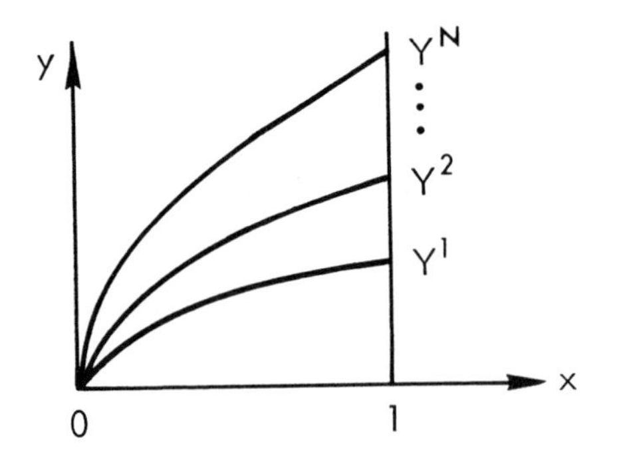

Figure 8-1
The Physical Situation

Fermat's principle of least time leads to the Euler equation [6] for a ray,

$$y'' = F(x,y,y',a,b), \tag{8.2a}$$

where

$$F(x,y,y',a,b) = n^{-1}\,[n_y - y'\,n_x][1 + y'^2]\ . \tag{8.2b}$$

The boundary conditions,

$$y\Big|_{x=0} = 0\ , \qquad y\Big|_{x=1} = Y^i\ , \quad i = 1,\ 2,\ \ldots,\ N, \tag{8.3}$$

together with Eqs. (8.2a and 8.2b), determine the functions

$$y = y(i,x,a,b), \quad i = 1,\ 2,\ \ldots,\ N, \tag{8.4}$$

which describe the paths of the N rays. The index i refers to the ith ray.

The time of arrival of a ray at the ith receiver is in theory given by the equation

$$T^i = T^i(a,b) = \int_0^1 n\left[1 + [y'(i,x,a,b)]^2\right]^{\frac{1}{2}} dx\ , \tag{8.5}$$

where

$$n = n(x,y(i,x,a,b),a,b). \tag{8.6}$$

The <u>observed</u> time $W^i(i = 1,\ 2,\ \ldots,\ N)$, however, may contain an error and is only approximately equal to the theoretical value,

$$W^i \cong T^i, \quad i = 1, 2, \ldots, N. \tag{8.7}$$

The problem is to find an optimal set of parameters (a,b). This set minimizes the function $S = S(a,b)$,

$$S(a,b) = \sum_{i=1}^{N} \lambda^i [T^i(a,b) - W^i]^2. \tag{8.8}$$

This quantity is a weighted sum of squares of deviations with weights $\lambda^i (i = 1, 2, \ldots, N)$.

8-3 QUASILINEARIZATION

Let us now digress to assure that we could numerically produce the solution of the nonlinear boundary-value problem of Eqs. (8.2) and (8.3) if a and b were known. We consider the determination of the ith path by the method of quasilinearization an extension of Newton's iteration method. Convergence is rapid, being quadratic in the limit.

Let $y_0(x) = y_0(i,x,a,b)$ be a current approximation to the path of a ray, and let $y_1(x) = y_1(i,x,a,b)$ denote the next approximation, which we seek. The differential equation, Eq. (8.2), is expanded as

$$y''_1 = F(x,y_0,y'_0,a,b) + (y_1 - y_0)F_y(x,y_0,y'_0,a,b)$$
$$+ (y'_1 - y'_0)F_{y'}(x,y_0,y'_0,a,b) \tag{8.9}$$
$$+ \ldots \ .$$

The higher order terms indicated by the dots are neglected, so that the equation becomes

$$y'_1 = F(x,y_0,y'_0,a,b) + (y_1 - y_0)F_y(x,y_0,y'_0,a,b)$$
$$+ (y'_1 - y'_0)F_{y'}(x,y_0,y'_0,a,b). \tag{8.10}$$

The subscripts y and y' on the function $F(x,y,y',a,b)$ denote partial differentiation with respect to the second and third arguments, respectively.

Equation (8.10) is a linear second-order differential equation for $y_1(x)$. The general solution is expressed, by the method of complementary solutions, as a linear combination of a particular solution and a complementary solution,

$$y_1(x) = p(x) + mh(x), \tag{8.11}$$

where m is a constant to be determined. The particular and complementary solutions satisfy the differential equations

$$p'' = F(x,y_0,y'_0,a,b) + (p - y_0)F_y(x,y_0,y'_0,a,b)$$
$$+ (p' - y'_0)F_{y'}(x,y_0,y'_0,a,b), \tag{8.12}$$

$$h'' = h\, F_y(x,y_0,y_0',a,b) + h'F_{y'}(x,y_0,y_0',a,b). \quad (8.13)$$

In order to meet the first boundary condition,

$$y_1(0) = 0, \quad (8.14)$$

we set

$$p(0) = 0, \quad (8.15)$$

$$h(0) = 0. \quad (8.16)$$

We choose a particular solution generated numerically and having some initial slope, say,

$$p'(0) = 0 \;. \quad (8.17)$$

The complementary solution is generated with

$$h'(0) = 1. \quad (8.18)$$

The constant multiplier m is determined so as to satisfy the second boundary condition

$$y_1(1) = Y^i \,, \quad (8.19)$$

or

$$p(1) + mh(1) = Y^i \,. \quad (8.20)$$

Since $p(1)$ and $h(1)$ are known numerically, this is an equation for the value of m in Eq. (8.11). The new approximation is then determined and the process may be repeated as necessary.

Let us summarize the steps involved in one cycle of quasilinearization on a computing machine. At the start we have stored the current approximation at a grid of points of step size Δ: $y_0(0)$, $y_0(\Delta)$, $y_0(2\Delta)$, ..., $y(1)$. The steps are

1. Produce the particular and complementary functions by numerical integration using Eqs. (8.12), (8.13), and (8.15) - (8.18), and store $p(0)$, $p(\Delta)$, ..., $p(1)$, $h(0)$, $h(\Delta)$, ..., $h(1)$.

2. Determine the multiplier m by solving Eq. (8.20).

3. Evaluate the new function $y(x)$ by using Eq. (8.11) and store $y_1(0)$, $y_1(\Delta)$, ..., $y_1(1)$ in place of y_0.

8-4 METHODS OF SOLUTION OF THE INVERSE PROBLEM

While we do not have the luxury of a closed-form solution of the inverse problem, there are several feasible successive approximation methods. We describe three methods, the third being the one for which we present numerical results.

1. Search Procedure: Produce the paths $y(i,x,\alpha,\beta)$ by quasilinearization as described in Section 8-2 and evaluate the sum $S(\alpha,\beta)$, using Eq. (8.8), for various allowed values of the parameters α and β. The optimal pair of a and b is that for which the quantity S is a minimum.

2. Gradient Method. The previous method is improved at the expense of calculating the first derivatives $S_a(\alpha,\beta)$ and $S_b(\alpha,\beta)$. From Eq. (8.8) and partial differentiation, the

expression for $S_a(\alpha,\beta)$ is

$$S_a(\alpha,\beta) = 2\sum_i \lambda^i \, [T^i(\alpha,\beta) - W^i] \, T_a^i(\alpha,\beta) \, . \tag{8.21}$$

From Eq. (8.5)

$$T_a^i(\alpha,\beta) = \int_0^1 n_a[1 + (y')^2]^{\frac{1}{2}} \, dx + \int_0^1 n[1 + (y')^2]^{-\frac{1}{2}} y' y'_a \, dx \, , \tag{8.22}$$

where all functions on the right side are evaluated at (i,x,α,β).

The first variation $y_a(x) = y_a(i,x,\alpha,\beta)$ is the solution of the linear two-point boundary-value problem

$$y''_a = F_y(x,y,y',\alpha,\beta)y_a + F_{y'}(x,y,y',\alpha,\beta)y'_a + F_a(x,y,y',\alpha,\beta), \tag{8.23}$$

$$y_a(i,0,\alpha,\beta) = 0, \qquad y_a(i,1,\alpha,\beta) = 0 \tag{8.24}$$

These equations are obtained by differentiation of Eqs. (8.2) and (8.3). Similar equations hold for $S_b(\alpha,\beta)$, $T_b^i(\alpha,\beta)$, and $y_b(x) = y_b(i,x,\alpha,\beta)$. They can be solved numerically as discussed above. Details of gradient methods for minimizing are given in Ref. 8.

3. Quadratically Convergent Method. To solve the nonlinear equations for minimizing S,

$$S_a(a,b) = 0, \qquad S_b(a,b) = 0 \, , \tag{8.25}$$

we use Newton's method. Expand the left sides and keep only the linear terms. The result is the system of linear equations for a and b

$$
\begin{aligned}
&S_a(\alpha,\beta) + (a-\alpha)S_{aa}(\alpha,\beta) + (b-\beta)S_{ab}(\alpha,\beta) = 0 \\
&S_b(\alpha,\beta) + (a-\alpha)S_{ba}(\alpha,\beta) + (b-\beta)S_{bb}(\alpha,\beta) = 0.
\end{aligned}
\tag{8.26}
$$

This method requires that the second derivatives of S be evaluated at $a = \alpha$ and $b = \beta$. Consider, for example, $S_{ab}(\alpha,\beta)$. From differentiation of Eq. (8.21) we have

$$
\begin{aligned}
S_{ab}(\alpha,\beta) = 2 \sum \lambda^i [&T^i_b(\alpha,\beta)\ T^i_a(\alpha,\beta) \\
&+ T^i(\alpha,\beta)\ T^i_{ab}(\alpha,\beta) - W^i\ T^i_{ab}(\alpha,\beta)] ,
\end{aligned}
\tag{8.27}
$$

and from Eq. (8.22) we have

$$
\begin{aligned}
T^i_{ab}(\alpha,\beta) = &\int_0^1 n_{ab}[1 + (y')^2]^{1/2}\, dx \\
&+ \int_0^1 n_a[1 + (y')^2]^{1/2}\, y'\, y'_b\, dx \\
&+ \int_0^1 n_b[1 + (y')^2]^{-1/2}\, y'\, y'_a\, dx \\
&- \int_0^1 n[1 + (y')^2]^{-3/2}\, y'\, y'_a\, y' y'_b\, dx \\
&+ \int_0^1 n[1 + (y')^2]^{-1/2}\, y'_b\, y'_a\, dx \\
&+ \int_0^1 n[1 + (y')^2]^{-1/2}\, y'\, y'_{ab}\, dx
\end{aligned}
\tag{8.28}
$$

Equations which determine $y_{ab}(x) = y_{ab}(i,x,\alpha,\beta)$ are obtained by differentiation of Eqs. (8.23) and (8.24),

$$y''_{ab} = F_y(x,y,y',\alpha,\beta)y_{ab} + F_{y'}(x,y,y',\alpha,\beta)y'_{ab} + F_{ab}(x,y,y',\alpha,\beta), \tag{8.29}$$

$$y_{ab}(i,0,\alpha,\beta) = 0, \qquad y_{ab}(i,1,\alpha,\beta) = 0 . \tag{8.30}$$

8-5 NUMERICAL METHOD AND RESULTS

Let us describe a numerical procedure for electronic computers based on the quadratic method. It is assumed that all functions are stored in the computer's memory as they are evaluated. To conserve storage locations, one ray at a time is considered. The steps involved in one cycle of the iterative process are as follows.

1. Using the current approximations, α and β, produce, for a given i, $y(i,x,\alpha,\beta)$ for $i = 1, 2, \ldots, N$, $0 < x < 1$, by solving the nonlinear two-point boundary-value problems of Eqs. (8.2) and (8.3) via quasilinearization.
2. Evaluate $T^i(\alpha,\beta)$ using Eq. (8.5).
3. Solve the linear boundary-value problems of Eqs. (8.23), (8.24), (8.29), and (8.30) for y'_a, y'_{ab}, and similar problems for the other derivatives.
4. Evaluate the quantities T^i_a, T^i_{ab}, ..., using Eqs. (8.22), (8.28), ...
5. Update the sums S_a, S_{ab}, ... , using Eqs. (8.21), (8.27), ...
6. After steps 1-5 are completed for all N rays,

solve Eqs. (8.26) for the new approximations, a and b.

In step 3, it is noted that there are five functions to be determined, y_a, y_b, y_{ab}, y_{aa}, and y_{bb}. The five boundary-value problems have exactly the same form (compare Eqs. (8.23) and (8.24) with (8.27) and (8.30); only the forcing functions differ. Hence, the task reduces to the production of one complementary solution and five particular solutions of initial-value problems.

We consider the example with index of refraction having the form

$$n(x,y,a,b) = 1 + a\, e^{-y} + bx, \tag{8.31}$$

where the true values of the parameters are

$$a = 0.1, \qquad b = 0.1\ . \tag{8.32}$$

The number of receivers is

$$N = 6 \tag{8.33}$$

and the ordinates of the receivers are

$$Y^1 = 0\ , \qquad Y^2 = 0.2, \ \ldots, \qquad Y^6 = 1.0. \tag{8.34}$$

The exact time of arrival of rays at the receivers computed by quasilinearization are given in Table 8-1.

The numerical procedure described above is applied to the problem of estimating the index of refraction profile from the data of Table 8-1. With a fourth-order Adams-Moulton method for numerical integration with a step size of

0.1, rapid convergence to the correct values of a and b is attained as shown in Table 8-2. Computing time is about half a minute on the IBM 7044. Note that the initial approximation is a homogeneous medium, but the estimation is quickly refined in about three iterations.

To test the effect of noisy data, we deliberately corrupt the exact data. The noisy time of arrival is

$$W_{noisy} = W_{exact} (1+pr) \tag{8.35}$$

where r is a random number generated with a standard deviation 1.0 and mean 0.0, and p is a decimal fraction. We perform 10 trials with $p = 0.001 = 0.1$ percent using 10 different sequences of random numbers. The 10 estimates of a and b are averaged. The mean values are $a \cong 0.096$, $b \cong 0.108$. The process is repeated with $p \cong 0.01 \cong 1.0$ percent, and results are $a \cong 0.059$, $b \cong 0.176$.

Table 8-1

Exact Times of Arrival

Ray Number	Time
1	1.14964
2	1.16277
3	1.21906
4	1.31141
5	1.43176
6	1.57293

Table 8-2

Convergence Using Exact Data

Approximation Number	Estimate of a	Estimate of b
Initial	0	0
1	0.1726	-0.0113
2	0.1078	0.0875
3	0.0998	0.1004
4	0.0994	0.1011
5	0.0994	0.1011

8-6 DISCUSSION

The numerical examples show that the quadratic method provides a feasible way of estimating the structure of a medium, even with noise present in the data. Other numerical experiments may be performed to assess the effect of noise on the estimates. Questions which still remain to be answered are how to use statistical information and how to use a priori information. Methods for making maximum likelihood estimates are fairly clear.

Other powerful methods are currently under investigation.

REFERENCES

1. Grant, F., and G. West, Interpretation Theory in Applied Geophysics, McGraw-Hill Book Company, New York, 1965.

2. Pedersen, M., and D. White, "Ray Theory of the General Epstein Profile," J. Acoust. Soc. Am., Vol. 44, Nol 3, 1968, pp. 765-786.

3. Bucker, H., and H. Morris, "Epstein Normal-Mode Model of a Surface Duct," J. Acoust. Soc. Am., Vol. 41, No. 6, 1967, pp. 1475-1478.

4. Bellman, R., D. Detchmendy, H. Kagiwada, and R. Kalaba, "On the Identification of Systems and the Unscrambling of Data-III: One-Dimensional Wave and Diffusion Processes," J. Math. Anal. Appl., Vol. 23, No. 1, July 1968, pp. 173-182.

5. Buell, J., H. Kagiwada, and R. Kalaba, "A Proposed Computational Method for Estimation of Orbital Elements, Drag Coefficients, and Potential Field Parameters from Satellite Measurements," Annales de Geophysique, Vol. 23, No. 1, January 1967, pp. 35-39.

6. Courant, R., and D. Hilbert, Methods of Mathematical Physics, Interscience Publishers, New York, Seventh Printing, 1966.

7. Mikhlin, S., and K. Smolitskiy, Approximate Methods for Solution of Differential and Integral Equations, American Elsevier Publishing Co., New York, 1967.

8. Tompkins, C. B., "Methods of Steep Descent," Modern Mathematics for the Engineer, E. Beckenbach (ed.), McGraw-Hill Book Company, Inc., New York, 1956.

Chapter 9

NONLINEAR FILTERING OF SIGNALS

9-1 INTRODUCTION

Recent years have seen the production of many significant advances in the theory of optimal filtering [1-7]. For the most part investigators have dealt with linear systems and have assumed that a complete statistical knowledge of the state of affairs is available. The aim of the present chapter is to introduce a theory which is applicable to a wide variety of nonlinear estimation problems. It is based on the theory of invariant imbedding [8-10], though no specialized knowledge on the part of the reader is assumed. Statistical assumptions are kept to a minimum, and there are immediate applications to problems of orbit determination [11] and identification problems in adaptive control.

9-2 FORMULATION

Let us consider a system which undergoes a process described by the nonlinear differential equation

$$\dot{x} = g(x,t). \tag{9.1}$$

We observe the process imperfectly; i.e., if

$$y(t) = \text{observed history of process on } 0 \leqq t \leqq T, \tag{9.2}$$

then

$$y(t) - x(t) = \text{observational error}. \tag{9.3}$$

Our aim is to make an estimate of the state of the system at time T on the basis of observations carried out on the interval $0 \leqq t \leqq T$. We shall make this estimate by selecting

$$x(T) = c \tag{9.4}$$

to be such that if $y(t)$ is the observed function, and the function $x(t)$ is determined on the interval $0 \leqq t \leqq T$ by the differential equation

$$\dot{x} = g(x,t) \tag{9.5}$$

and the condition

$$x(T) = c, \tag{9.6}$$

then the integral

$$J = \int_0^T (x(t) - y(t))^2 dt \tag{9.7}$$

is minimized.

9-3 INVARIANT IMBEDDING

Let us recognize that the value of the integral in Eq. (9.7) is a function of T, the length of the time of observation, and c, the value assigned to $x(t)$ at time T,

$$\int_0^T (x(t) - y(t))^2 dt = f(c,T). \tag{9.8}$$

Now we imbed the original process, involving fixed values of T and c, in a class of processes for which $0 \leqq T$ and $-\infty < c < +\infty$. Then we interconnect the costs $f(c,T)$, for these processes. We see that

$$f(c,T + \Delta) = f(c - g(c,T)\Delta, T) + (y(T) - c)^2\Delta + o(\Delta), \tag{9.9}$$

or in the limit as $\Delta \to 0$,

$$f_T = (y(T) - c)^2 - g(c,T)f_c, \tag{9.10}$$

which is a linear first-order partial differential equation for the function $f(c,T)$.

To obtain an auxiliary condition on the function $f(c,T)$, we consider $f(c,0)$, which is the cost associated with estimating

$$x(0) = c \tag{9.11}$$

solely on the basis of the a priori information available. We might select the condition

$$f(c,0) = k(c - c_0)^2, \tag{9.12}$$

where c_0 is our best estimate of the quantity $x(0)$, and k is a measure of our confidence in that estimate. The linear partial differential Eq. (9.10) together with the condition in Eq. (9.12) determine the function $f(c,T)$.

Assume that we have been observing the process in the interval $0 \leqq t \leqq T_0$. Our estimate of the current state is the value of c which minimizes the cost function $f(c,T_0)$. In fact, we see that we would like to determine the minimizing value of c for each value of $T \geqq 0$. These are our estimates, $e(T)$. These minimizing points satisfy the equation

$$f_c(e,T) = 0 , \tag{9.13}$$

or

$$f_{cc}(e,T)de + f_{cT}(c,T)dT = 0 , \tag{9.14}$$

$$\frac{de}{dT} = - f_{cT}(e,T)/f_{cc}(e,T) . \tag{9.15}$$

Once the function $y(t)$ has been observed on the interval $0 \leqq t \leqq T$, and the functions $f(e,T)$, $f_{cT}(e,T)$ and $f_{cc}(e,T)$ have been determined, the Eq. (9.15) provides the optimal estimate of $x(T)$, $e(T)$. As an initial condition we might use the condition

$$e(0) = c_0 . \tag{9.16}$$

9-4 CONTINUATION OF THE ANALYSIS

Fortunately, we can continue our analysis much further. We take the partial derivatives of both sides of Eq. (9.10) to obtain the equation

$$f_{Tc} + g_c f_c + g f_{cc} = -2(y(T) - c) . \tag{9.17}$$

This yields

$$-f_{cT}/f_{cc} = (g_c f_c/f_{cc}) + g + (2(y - c))/f_{cc} . \tag{9.18}$$

Since

$$f_c(e,T) = 0 , \tag{9.19}$$

we see that Eq. (9.15) becomes

$$\frac{de}{dT} = g(e,T) + (2/f_{cc}(e,T))(y-e) . \tag{9.20}$$

If we introduce

$$2/f_{cc}(e(T),T) = q(T) , \tag{9.21}$$

we see that Eq. (9.20) assumes a remarkable form

$$\frac{de}{dT} = g(e,T) + q(T)(y-e) . \tag{9.22}$$

It states that the optimal estimate of the current state, $e(T)$, is obtained by integrating a slightly modified form of the original system equation

$$\dot{x} = g(x,t), \tag{9.23}$$

the modification being a term $q(T)\ (y-e)$, where $y - e$ is the difference between the observed and estimated values and q is a weighting factor. This is an extension of well-known results for the linear case. [5]

9-5 EQUATION FOR THE WEIGHTING FACTOR

To derive an equation for the weighting factor

$$q(T) = 2/f_{cc}(e(T),T) \tag{9.24}$$

we differentiate Eq. (9.17) with respect to c, which yields the relation

$$(f_{cc})_T + g(f_{cc})_c = 2[1 - g_c f_{cc}] - g_{cc} f_c . \tag{9.25}$$

Also we have

$$\frac{dq}{dT} = - 2f_{cc}^{-2}[f_{ccc} \frac{de}{dT} + f_{ccT}]. \tag{9.26}$$

We can use Eqs. (9.24), (9.25), (9.26), and (9.22) to rewrite this equation in the form

$$\frac{dq}{dT} = 2(q^2/4)[f_{ccc}\{g + q(y - e)\}$$

$$+ 2\{1 - g_c(2/q)\} - gf_{ccc}], \tag{9.27}$$

$$\frac{dq}{dT} = 2g_c\, q - q^2 - \frac{q^3}{2}\, f_{ccc}(y - e). \tag{9.28}$$

We can again derive an equation for f_{ccc} by noting that

$$\frac{d}{dT}\, f_{ccc}(e(T),T) = f_{cccc} \cdot \frac{de}{dT} + f_{cccT} \cdot \tag{9.29}$$

Differentiating Eq. (9.25) with respect to c yields

$$f_{cccT} = g_c f_{ccc} + g f_{cccc} = -2\{g_c f_{ccc} + g_{cc} f_{cc}\}$$
$$- g_{ccc} f_c - g_{cc} f_{cc} \,. \tag{9.30}$$

Combining Eq. (9.29) and (9.30) yields

$$\frac{d}{dT}\, f_{ccc} = f_{cccc}\left\{\frac{de}{dT} - g\right\} - 3g_c f_{ccc} - 3g_{cc} f_{cc}$$
$$= -6\,\frac{g_{cc}}{q} - 3\, g_c f_{ccc} + q(T)\, f_{cccc}\,(y-e)\,. \tag{9.31}$$

9-6 PRACTICAL CONSIDERATIONS

We note that Eq. (9.31) involves an additional unknown variable f_{cccc}. We can, of course, derive an additional differential equation for f_{cccc} which will again involve f_{ccccc}.

In practice, this refinement is quite unnecessary. We assert that the function $f(c,T)$ is representable in the form $a_0(T) = a_1(T)c + a_2(T)c^2$ in the neighborhood of the optimal estimate $e(T)$. This implies that f_{ccc} is negligible in the neighborhood of the optimal estimate.

With this assumption, the estimator equations are, from Eqs. (9.22) and (9.28),

$$\frac{de}{dT} = g(e,T) + q(T)(y - e) \tag{9.32}$$

and

$$\frac{dq}{dT} = 2g_c(e,T)\, q - q^2. \tag{9.33}$$

9-7 NUMERICAL RESULTS

To test the method we did some numerical experiments on a digital computer. We considered a system described by the nonlinear differential equation and initial condition

$$\dot{x} = -x + \varepsilon x^3/3, \quad 0 \leq t \leq 5 , \tag{9.34}$$

$$x(0) = x_o. \tag{9.35}$$

We put $x_o = 1$ and $\varepsilon = 0.1$ and produced the solution, which is shown in Fig. 9-1. Next we produced values of

$$y(t) = x(t) + .5 \cos(60t) , \tag{9.36}$$

which is to serve as our "noisy" observation of the process $x(t)$. Then we tried several experiments to see how well and how rapidly we could determine the true curve $x(t)$ from the noisy observations. The equations being integrated are, according to Eqs. (9.32) and (9.33),

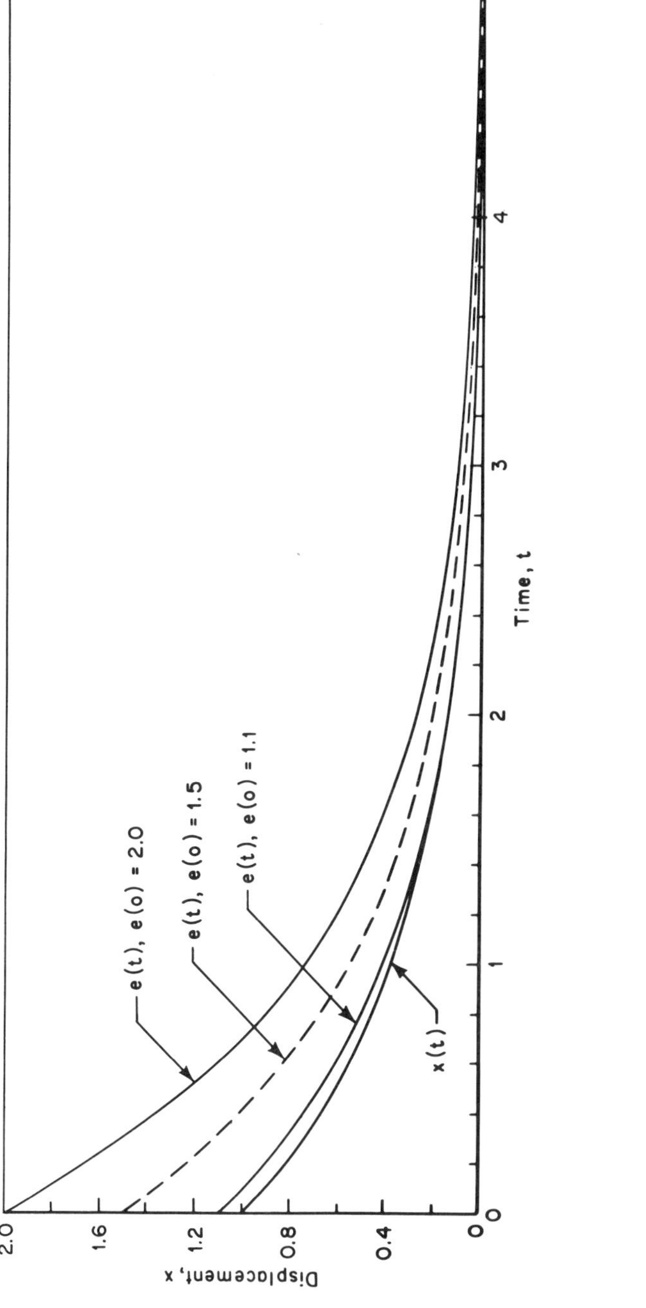
e(t), e(o) = 2.0
e(t), e(o) = 1.5
e(t), e(o) = 1.1
x(t)
Displacement, x
2.0
1.6
1.2
0.8
0.4
0
Time, t
0
1
2
3
4
5

Figure 9-1

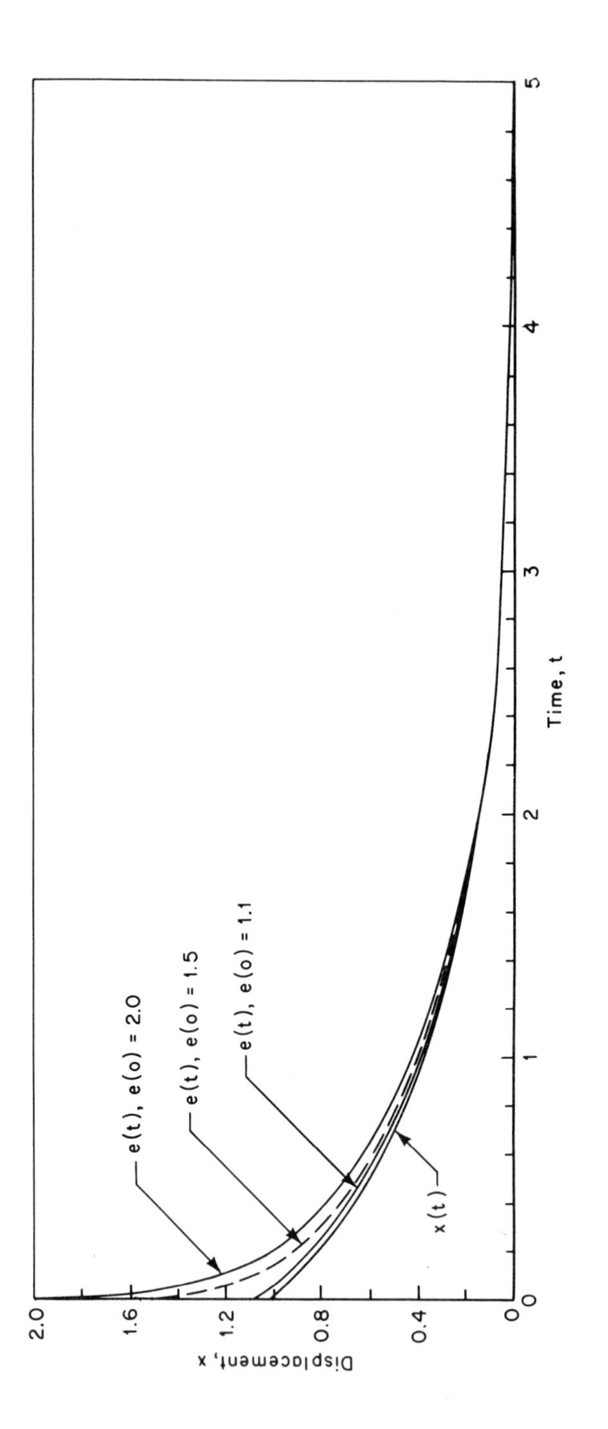
Displacement, x
Time, t
e(t), e(o) = 2.0
e(t), e(o) = 1.5
e(t), e(o) = 1.1
x(t)
2.0
1.6
1.2
0.8
0.4
0
1
2
3
4
5

Figure 9-2

$$\frac{de}{dT} = -e + \varepsilon\frac{e^3}{3} + q[y(T) - e], \quad e(0) = e_0, \tag{9.37}$$

$$\frac{dq}{dT} = 2[\varepsilon e^2 - 1]\, q - q^2, \quad q(0) = q_0 \; . \tag{9.38}$$

In the first set of experiments we put $q_0 = 0.2$ and tried three initial values for $e(0)$,

$$e(0) = 2\ x(0) = 2 \ ,$$

$$e(0) = y(0) = 1.5 \ ,$$

$$e(0) = 1.1\ x(0) = 1.1 \ .$$

In all cases by time $t = 5$ the observations, represented by $y = y(t)$, have been filtered and $e(t) \cong x(t)$. Then we tried the same experiment but with $q_0 = q(0) = 20$. In this case excellent filtering is accomplished by time $t = 2$, as is shown in Fig. 9-2.

9-8 SYSTEM IDENTIFICATION

The approach presented here appears rather promising in many applications. As an example we cite the identification problem in adaptive control where one has to determine a running (on-line) estimate of certain plant parameters based on input-output observations before exerting further control. When suitably formulated, this problem takes the form:

$$\dot{\underline{x}} = \underline{f}(t, \underline{x}, \underline{u}, \underline{a}), \tag{9.39}$$

where $\underline{x}$ is the state vector, $\underline{u}$ the control vector and $\underline{a}$ is an unknown parameter vector, determine the parameter vector $\underline{a}$ based on observations

$$\underline{y}(t) = \underline{g}(t,x) + \text{observational error} \tag{9.40}$$

in order to minimize

$$\int_0^T \| \underline{y}(t) - \underline{g}(t,x) \|_Q \, dt \, , \tag{9.41}$$

where $\|\cdot\|_Q$ denotes a suitable semi-norm.

We note that this problem is exactly in the framework proposed in this note as soon as we adjoin the additional equation

$$\dot{\underline{a}} = 0. \tag{9.42}$$

It would be interesting to find out if it will be possible to extend this approach when Eq. (9.1) involves a random forcing term.

9-9 NUMERICAL RESULTS

To test this method we considered a system described by the nonlinear differential equation

$$\ddot{x} + 3\dot{x} + 2x + ax^3 = u(t), \tag{9.43}$$

which corresponds to a spring, mass and dashpot system with a nonlinear spring.

The uncorrupted data were generated with initial conditions

$$x(0) = 2, \ \dot{x}(0) = 0 , \tag{9.44}$$

with the value of the parameter a being

$$a = 0.5 , \tag{9.45}$$

and input

$$u(t) = 5 \sin t. \tag{9.46}$$

We produced values of

$$y(t) = x(t) + \eta(t), \tag{9.47}$$

which is to serve as our "noisy" observation of $x(t)$.

The noise model for $\eta(t)$ was of the form

$$\eta(t) = 0.1 \, r_1(t) |x(t)| + 0.1 \, r_2(t) , \tag{9.48}$$

where $r_1(t)$ and $r_2(t)$ for each t were random variables with a uniform distribution in the interval $(-1, 1)$.

Based on the noisy observation $y(t)$ of equation (9.47), we estimated $x(t)$, $\dot{x}(t)$ and a.

To do this, the system equations are written in the form

$$\begin{aligned} \dot{x}_1 &= x_2 \\ \dot{x}_2 &= -2x_1 - ax_1^3 - 3x_2 + u(t) \\ \dot{a} &= 0, \end{aligned} \tag{9.49}$$

where $x_1 = x(t)$ and $x_2 = \dot{x}(t)$.

Denoting the estimates of x_1, x_2, and a by e_1, e_2, and e_3 respectively, it is seen that the estimator equations are

$$\frac{d}{dt}\begin{bmatrix} e_1 \\ e_2 \\ e_3 \end{bmatrix} = \begin{bmatrix} e_2 \\ -2e_1 - e_3 e_1^3 - 3e_2 + u \\ 0 \end{bmatrix} + q \begin{bmatrix} 1 \\ 0 \\ 0 \end{bmatrix} (y - e_1) \tag{9.50}$$

and

$$\frac{dq}{dT} = \frac{1}{2}\left[g_c q + q g_c^T\right] - q^2 \tag{9.51}$$

where

$$q = (q_{ij}), \quad i, j = 1, 2, 3 \tag{9.52}$$

and

$$g_c = \begin{bmatrix} 0 & -2-3e_3 e_1^2 & 0 \\ 1 & -3 & 0 \\ 0 & -e_1^3 & 0 \end{bmatrix}. \tag{9.53}$$

The following initial conditions were used for the estimator

$$\begin{bmatrix} e_1(0) \\ e_2(0) \\ e_3(0) \end{bmatrix} = \begin{bmatrix} y(0) \\ 0.5 \\ 0 \end{bmatrix} \tag{9.54}$$

and

$$q(0) = \begin{bmatrix} 1 & 0 & 0 \\ 0 & 3 & 0 \\ 0 & 0 & 5 \end{bmatrix}. \tag{9.55}$$

The results of the experiment are shown in Figures 9-3, 9-4, and 9-5. It is seen that both the estimate of the parameter and the tracking are excellent after 2 seconds.

9-10 SUMMARY AND DISCUSSION

We have considered a system undergoing a process described by the differential equation

$$\dot{x} = g(x,t).$$

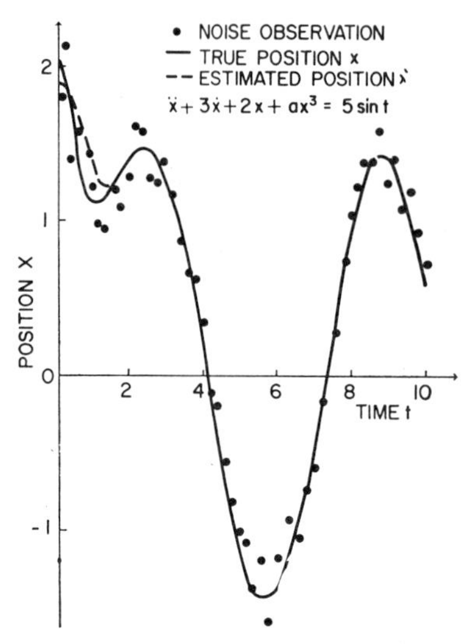

Figure 9-3

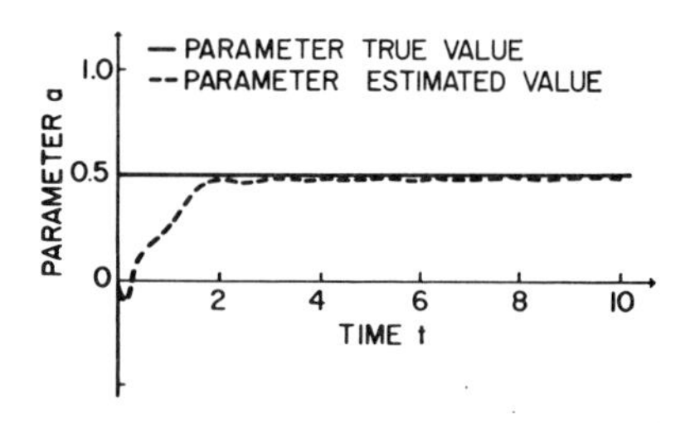

Figure 9-5

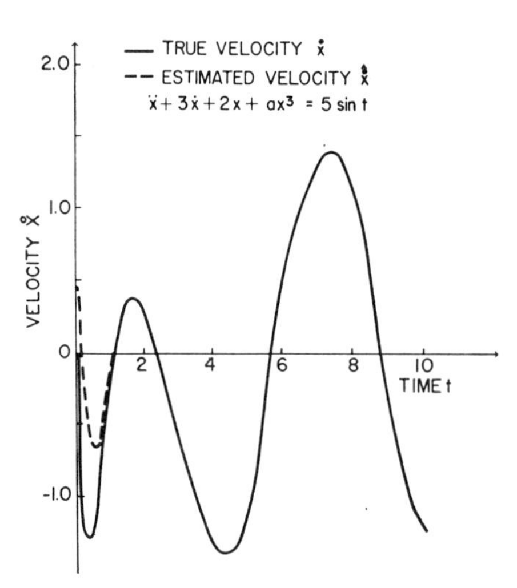

Figure 9-4

On the time interval $(0,T)$ we observed the function x, in a noisy manner, and denoted this experimental function by the symbol y. We showed how to determine the state of the system at time $t = T$ such that J,

$$J = \int_0^T (x(t)-y(t))^2 dt$$

is minimized. These concepts were based in part on the theory of invariant imbedding. Some numerical experiments were reported. We further showed how to extend the method to estimate system parameters.

In this chapter, we have not taken up the interpolation problem in which one is interested in obtaining an estimate of the states and parameters of the system at a finite set of fixed instants of time which are contained in the interval of observation $(0,T)$. This is done in Chapter 10.

REFERENCES

1. Wiener, N., Extrapolation, Interpolation and Smoothing of Stationary Time Series, John Wiley and Sons, Inc., New York, 1950.

2. Middleton, D., An Introduction to Statistical Communication Theory, McGraw-Hill Book Company, Inc., New York, 1960.

3. Swerling, P., "First Order Error Propagation in a Stagewise Smoothing Procedure for Satellite Observations," Journ. Astronautical Sciences, Vol. 6, No. 3, Autumn 1959, pp. 46-52.

4. Pugachev, V., Theory of Random Functions and its Application to Automatic Control Problems (in Russian), Gostekhizdat, Moscow, 1960.

5. Kalman, R., and R. Bucy, "New Results in Linear Filtering and Prediction Theory," Jour. of Basic Engineering, Trans. ASME, Series D, Vol. 83, No. 1, March 1961, pp. 95-108.

6. Ho, Y., The Method of Least Squares and Optimal Filtering Theory, The Rand Corporation, RM-3329-PR, October 1962.

7. Cox, H., "Estimation of State Variables," Abstract No. M 2564 in Progress Report Number 13 of the Research and Educational Activities in Machine Computation by the Cooperating Colleges of New England, Computation Center, Massachusetts Institute of Technology, July 1963, p. 163.

8. Bellman, R.E., R. E. Kalaba, and G. M. Wing, "Invariant Imbedding and Mathematical Physics I. Particle Processes," Jour. of Mathematical Physics, Vol. 1, No. 4 (1960) pp. 280-308.

9. Bellman, R. E., H. H. Kagiwada, R. E. Kalaba, and M. C. Prestrud, Invariant Imbedding and Time-dependent Transport Processes, American Elsevier Publishing Company, Inc., New York, 1964.

10. Kalaba, R.E., "Invariant Imbedding and the Analysis of Processes," Chapter 10 of the book Views on General Systems Theory, edited by M. Mesarivoc, John Wiley and Sons, Inc., New York, 1964.

11. Bellman, R. E., H. H. Kagiwada, and R. E. Kalaba, "Orbit Determination as a Multi-point Boundary-value Problem and Quasilinearization," Proc. Nat. Acad. Sci. USA, Vol. 48 (1962), pp. 1327-1329.

Chapter 10

NONLINEAR INTERPOLATING FILTER FOR IMPRECISE DYNAMICAL EQUATIONS

10-1 INTRODUCTION

The estimation of states and parameters in noisy nonlinear dynamic systems based on observations on noisy nonlinear combinations of some or all of the states is a problem whose solution is frequently important. Depending on the instant at which the estimate is desired, the problem can be associated with estimation at a future time, current time or past time (interpolation problem). For control applications, sequential solutions of these problems are often desired.

For problems with a linear structure, Kalman and Bucy [1] have obtained sequential solutions for estimating at a future and current time for a wide variety of statistical cost functions. Bellman, et al. [2] and Detchmendy and Sridhar [3] have obtained sequential solutions for a class of nonlinear problems with a least-squares cost function when the estimate is desired at the current time.

Here, we shall find a sequential solution for the nonlinear interpolation problem with a least-squares cost function. In the interpolation problem, one is primarily interested in obtaining an estimate of states and parameters of the system at a finite set of fixed instants of time which are contained in the interval of observation. The equations associated with the sequential solution will be called the equations of the sequential interpolating filter. We permit inaccuracies in the dynamical equations themselves, which is important in many applications.

This work is pertinent to problems of astronomy in which one wishes to update the estimates of initial conditions as additional observations become available, and to trajectory analysis problems in which one may wish to improve estimates at "epochs." Typically these epochs are associated with mid course correction times, instants at which photographs were taken, etc.

In this paper the detailed formulation and solution of the sequential interpolating problem is given for the scalar case to avoid unnecessary matrix manipulations. The generalization to the vector case is indicated briefly.

10-2 FORMULATION OF THE PROBLEM

The process to be estimated is described by

$$\dot{x} = g(t,x) + (\text{dynamical error}) \tag{10.1}$$

and observations on the process are made in the form

$$y(t) = h(t,x) + \text{(observational error)} \tag{10.2}$$

over the interval $0 \leq t \leq T$. It is required to estimate the state of the process at a fixed instant t_1, where $0 \leq t_1 \leq T$, based on knowledge of $y(t)$ in the interval $0 \leq t \leq T$. The generalization of this problem when the state of the process has to be estimated at a finite set of fixed instants of time $t_1, t_2, \ldots, t_n$, where $0 \leq t_1 \leq t_2 \leq \ldots \leq t_n \leq T$ is straightforward and will not be elaborated here.

In equations (10.1) and (10.2) the dynamic and observational error terms take into account the imprecise knowledge of the righthand sides of the equations, as well as any disturbances.

For any guessed estimate $\bar{x}(t)$ defined in the interval $0 \leq t \leq T$, an associated cost is defined by the functional

$$k_3\{\bar{x}(0) - m_0\}^2 + \int_0^T [k_1\{y(t) - h(t,\bar{x})\}^2 + k_2\{\dot{\bar{x}} - g(t,\bar{x})\}^2]dt \tag{10.3}$$

where k_1 is a nonnegative function of time, k_2 is a positive function of time defined in the interval $0 \leq t \leq T$, and k_3 is a nonnegative constant. The quantities m_0 and k_3 are related to prior knowledge about the state before any observations are available.

The optimal estimate $x^*(t)$ minimizes the functional (3). The optimal estimate at the intermediate time t_1 is then given by $x^*(t_1)$.

The determination of the optimal estimate is clearly equivalent to the following optimal control problem. (The bars in equation 10.3 will be suppressed henceforth for convenience.) Determine the optimal "control" $u(t), 0 \leq t \leq T$ and the corresponding optimal trajectory $x(t)$ to minimize

$$I[u(t)] = \int_0^T \left[k_1\{y(t) - h(t,x)^2\} + k_2 u^2\right] dt + k_3\{x(0) - m_0\}^2 \tag{10.4}$$

subject to the differential constraint

$$\dot{x} = g(t,x) + u \tag{10.5}$$

with $x(0)$ and $x(T)$ being free.

The value of the optimal trajectory at time t_1 is then the optimal estimate of interest.

10-3 A TWO-POINT BOUNDARY VALUE PROBLEM

From well-known results in optimal control theory [4] the canonic equations to be satisfied along an optimal trajectory are given by

$$\dot{x} = g(t,x) - \frac{1}{2k_2}\lambda \tag{10.6}$$

$$\dot{\lambda} = 2k_1\{y(t) - h(t,x)\}h_x(t,x) - g_x(t,x)\lambda . \tag{10.7}$$

From the transversality conditions, the boundary conditions

are

$$\lambda(0) = 2k_3(x(0) - m_0),$$
$$\lambda(T) = 0 . \qquad (10.8)$$

The solution of the two-point boundary value problem for a fixed value T will yield the optimal trajectory $x^*(t)$ and the corresponding estimate $x^*(t_1)$ for the given value T. The determination of the dependence of the optimal estimate $x^*(t_1)$ on T affords a means for sequential estimation. With this in view, we will make usc of the technique of invariant imbedding [5,6,7].

10-4 THE EQUATIONS OF INVARIANT IMBEDDING

Consider now the solution of the following differential equations

$$\dot{x} = \alpha(t,x,\lambda), \qquad (10.9a)$$

$$\dot{\lambda} = \beta(t,x,\lambda), \qquad (10.9b)$$

with the boundary conditions

$$\lambda(0) = 2k_3(x(0) - m_0),$$
$$\lambda(T) = c \qquad (10.10)$$

where

$$T \geq 0, \quad -\infty < c < \infty .$$

Let the corresponding value of the solution $x(t)$ at the instant t_1, be denoted by $\phi(t_1,T,c)$ where $0 \le t_1 \le T$.

Later we shall identify the right-hand sides of equations (10.6) and (10.7) with the right-hand sides of equation (10.9). In this case, it is clear that the optimal estimate is $\phi(t_1,T,0)$.

From equation (10.9b), it is evident that

$$\begin{aligned} \lambda(T-\Delta) &= \lambda(T) - \Delta\dot{\lambda}(T) + 0(\Delta^2) \\ &= c - \Delta\beta(T,\phi(T,T,c),c) + 0(\Delta^2). \end{aligned} \tag{10.11}$$

For $\lambda(T-\Delta)$ satisfying equation (10.11), clearly

$$\phi(t_1,T-\Delta,\lambda(T-\Delta)) = \phi(t_1,T,c) \ . \tag{10.12}$$

From equations (10.11) and (10.12),

$$\begin{aligned} \phi(t_1,T,c) = \phi(t_1,&T-\Delta,c-\Delta\beta(T,\phi(T,T,c)) \\ &+ 0(\Delta^2)) \ . \end{aligned} \tag{10.13}$$

Expanding the right-hand side of equation (10.13) about (t_1,T,c), canceling the appropriate terms, dividing throughout by Δ, and taking the limit as $\Delta \to 0$ yields the invariant imbedding equation for $\phi(t_1,T,c)$ in the form

$$\phi_T(t_1,T,c) + \beta(T,\psi(T,c),c)\ \phi_c(t_1,T,c) = 0 \tag{10.14}$$

where

$$\psi(T,c) = \phi(T,T,c) \ . \tag{10.15}$$

We now need to determine an equation satisfied by $\psi(T,c)$. Physically, $\psi(T,c)$ is the missing terminal condition on x for the two-point boundary-value problem. It is evident from equation (10.9a) that, for all $\lambda(T-\Delta)$ satisfying equation (10.11),

$$\psi(T-\Delta, \lambda(T-\Delta)) = \psi(T,c) - \Delta\alpha(T,\psi(T,c),c) + 0(\Delta^2) . \tag{10.16}$$

Substituting for $\lambda(T-\Delta)$ from equation (10.11) in (10.16), expanding the left-hand side about (T,c), canceling the appropriate terms, dividing throughout by Δ, and taking the limit as $\Delta \to 0$, the invariant imbedding equation for $\psi(T,c)$ is obtained in the form

$$\psi_T(T,c) + \beta(T,\psi(T,c),c)\, \psi_c(T,c) = \alpha(T,\psi(T,c),c). \tag{10.17}$$

It is evident from equation (10.10) that

$$c = 2k_3[\phi(0,0,c) - m_0]$$
$$= 2k_3[\psi(0,c) - m_0]$$

so that

$$\psi(0,c) = \frac{1}{2k_3} c + m_0 . \tag{10.18}$$

Equations (10.17) and (10.18) now constitute an initial-value problem. In principle, this is solved up to the fixed instant of time t_1. At this time, the initial condition

$$\phi(t_1,t_1,c) = \psi(t_1,c) \tag{10.19}$$

is imposed on equation (10.14), and equations (10.14) and (10.17) are simultaneously solved. This produces $\phi(t_1,T,c)$ for fixed t_1. The quantity $\phi(t_1,T,0)$, which is the estimate at time t_1, is now readily obtained. Thus equations (10.17) and (10.14) are the equations of the sequential interpolating filter. The filter equations are not in a very convenient form from the practical point of view.

10-5 A LINEAR SYSTEM

We will now carry out the steps outlined earlier for a linear problem. Specifically, let $g(t,x)$ and $h(t,x)$ of equations (10.1) and (10.2) take the form

$$\begin{aligned} g(t,x) &= g(t)x + f(t), \\ h(t,x) &= h(t)x . \end{aligned} \tag{10.20}$$

Comparing the right-hand sides of equations (10.6), (10.7), and (10.9) yields, from equation (10.20),

$$\begin{aligned} \alpha(t,x,\lambda) &= g(t)x + f(t) - \frac{1}{2k_2}\lambda , \\ \beta(t,x,\lambda) &= 2k_1\{y(t) - h(t)x\}h(t) - g(t)\lambda \end{aligned} \tag{10.21}$$

Substituting from equation (10.21) into equation (10.17), assuming a solution to the resulting equation of the form

$$\psi(T,c) = p(T)c + q(T) \tag{10.22}$$

and equating coefficients of like powers of c on both sides yields

$$\dot{p}(T) = 2k_1 h^2(T)p^2(T) + 2g(T)p(T) - \frac{1}{2k_2}, \tag{10.23}$$

$$\dot{q}(T) = g(T)q(T) = f(T) - 2k_1 h(T)p(T)\{y(T) - h(T)q(T)\} . \tag{10.24}$$

From equations (10.18) and (10.22)

$$p(0) = \frac{1}{2k_3} , \tag{10.25}$$

$$q(0) = m_0 .$$

Note that equation (10.23) is not forced by the observation. It can be solved with the suitable initial condition without solving for $q(T)$. Assume now a solution for equation (10.14) of the form

$$\phi(t_1,T,c) = R(t_1,T)c + r(t_1,T) . \tag{10.26}$$

Again, substituting from equations (10.26) and (10.21) into equation (10.14) and equating coefficients of like powers of c yields

$$R_T(t_1,T) = g(T)R(t_1,T) + 2k_2 h^2(T)R(T,T)$$

$$r_T(t_1,T) = -2k_1 h(T)R(t_1,T)\{y(T) - h(T)r(T,T)\} . \tag{10.27}$$

From equations (10.15), (10.22), and (10.26), it follows that

$$\begin{aligned} R(T,T) &= p(T) , \\ r(T,T) &= q(T) . \end{aligned} \tag{10.28}$$

From equations (10.27) and (10.28)

$$\begin{aligned} R_T(t_1,T) &= g(T)R(t_1,T) + 2k_1 h^2(T)p(T) , \\ r_T(t_1,T) &= - 2k_1 h(T)R(T_1,T)\{y(T) - h(T)q(T)\}. \end{aligned} \tag{10.29}$$

From equation (10.19), it follows that

$$\begin{aligned} R(t_1,t_1) &= p(t_1), \\ r(t_1,t_1) &= q(t_1) . \end{aligned} \tag{10.30}$$

The derivations are now complete. The sequential estimator can now be easily implemented.

Equations (10.23) and (10.24) are solved with the initial conditions (10.25) till the fixed instant t_1. At this instant equation (10.29) is adjoined to equations (10.23) and (10.24) and solved with the condition (10.30).

Since the optimal estimate at time t_1 is $\phi(t_1,T,0)$, it follows from equation (10.24) that $r(t_1,T)$ is indeed the estimate that we are seeking. Thus equations (10.23), (10.24), and (10.29) are the sequential interpolating filter equations.

10-6 AN APPROXIMATE SOLUTION FOR THE NONLINEAR PROBLEM

Since we are only interested in determining solutions of equations (10.14) and (10.17) in the neighborhood of $c = 0$ to obtain estimates, it is possible to obtain approximate solutions by simple expansion techniques. This is very similar to the scheme used for obtaining approximate solutions for the filtering problem in references [2 and 3]. The one-term expansion which seems to be quite adequate for all the test problems considered in the references will be considered next.

Assume a solution of the form given by equation (10.22) for equation (10.17), and equate coefficients of like powers of c. This will yield the approximate solution of equation (10.17). It is convenient here to note that

$$\begin{aligned}\beta(T,\psi(T,c),c) &= \beta(T,\, p(T)c+q(T),\, c)\\ &= \beta(T,q(T),0) + \beta_x(T,q(T),0)\, p(T)\, c \\ &\quad + \beta_\lambda(T,q(T),0)c + 0(c^2)\end{aligned} \tag{10.31}$$

and

$$\alpha(T,\psi(T,c),c) = \alpha(T,q(T),0) + \{\alpha_x(T,q(T),0) + \alpha_\lambda(T,q(T),0)\}c + 0(c^2) \ . \qquad (10.32)$$

The approximate solution is now easily seen to be

$$\dot{q}(T) = \alpha(T,q(T),0) - p(T)\beta(T,q(T),0) \qquad (10.33)$$

$$\dot{p}(T) = \alpha_x(T,q(T),0) + \alpha_\lambda(T,q(T),0) - p(T)[\beta_x(T,q(T),0)\ p(T) + \beta_\lambda(T,q(T),0)]. \qquad (10.34)$$

From equation (10.18),

$$q(0) = m_0 \ , \qquad p(0) = \frac{1}{2k_3} \ . \qquad (10.35)$$

Now, assume an approximate solution for equation (10.14) of the form of equation (10.26). Expand suitably and equate coefficients of like powers of c. After a slight arrangement, this yields

$$r_T(t_1,T) = -R(t_1,T)\beta(T,q(T),0),$$
$$R_T(t_1,T) = -R(t_1,T)[\beta_x(T,q(T),0)\ p(T) + \beta_\lambda(T,q(T),0)] \ . \qquad (10.36)$$

From equation (10.19), we get

$$r(t_1,t_1) = q(t_1), \qquad R(t_1,t_1) = p(t_1) \ . \qquad (10.37)$$

Now identifying α and β with the right-hand sides of equation (10.6) and equations (10.7), equations (10.33) through (10.37) can be rewritten as follows:

$$\dot{q}(T) = g(T,q(T)) - 2k_1 p(T)\{y(T) - h(T,q(T))\} h_x(T,q(T)) \quad (10.38)$$

$$\dot{p}(T) = 2g_x(T,q(T))p(T) - p^2(T)[2k_1 h_{xx}(T,q(T))\{y(t) - h(T,q(T))\} - 2k_1 h_x^2(T,q(T))] - \frac{1}{2k_2} \quad (10.39)$$

$$r_T(t_1,T) = -2k_1 R(t_1,T)\{y(T) - h(T,q(T))\}h_x(t,q(T))$$

$$R_T(t_1,T) = - R(t_1,T) \; [\{2k_1 h_{xx}(T,q(T))(y(T) - h(T,q(T))) - 2k_1 h_x^2(T,q(T))\}p(T) - g_x(T,q(T))] \; . \quad (10.41)$$

Equations (10.38) through (10.41), together with the initial conditions (10.35) and (10.37), constitute the equations of the approximate interpolating filter.

10-7 CONCLUSIONS

The solution to the sequential interpolating problem has been obtained by converting an associated boundary-value problem to an initial-value problem using the method of invariant imbedding. The key to sequential solution is that this does not involve associated problems which allow only iterative solutions. It appears that invariant imbedding techniques can be effectively utilized to obtain meaningful solutions to a variety of control and estimation problems.

Subsequent numerical work has demonstrated the effectiveness of this method.

APPENDIX

10-8 VECTOR GENERALIZATION

The analogs of the equations in the prior development for the scalar problem will be written down for the vector problem. The explanations are not elaborated in this section. The process to be estimated is described by

$$\dot{x} = g(t,x) + (\text{dynamical error}) \tag{10.42a}$$

where

$$x = \mathrm{col}(x_1, \ldots, x_n)$$

$$g = \mathrm{col}(g_1, \ldots, g_n)$$

and observations on the process are made in the form

$$y(t) = h(t,x) + (\text{observational error}) \tag{10.42b}$$

where

$$y = \mathrm{col}(y_1, \ldots, y_m)$$

$$h = \mathrm{col}(h_1, \ldots, h_m) \cdot$$

The cost associated with the estimate is given by

$$||\bar{x}(0) - m_0||^2_{k_3} + \int_0^T \Big[||y(t) - h(t,\bar{x})||^2_{k_1} + ||\dot{x} - g(t,\bar{x})||^2_{k_2}\Big]\, dt \qquad (10.43)$$

where k_1 and k_3 are symmetric matrices of appropriate dimensions which are at least positive semi-definite and k_2 is a symmetric matrix which is positive definite. The symbol $||z||^2_M = (z,Mz)$ where z is a p vector and M is a $p \times p$ symmetric matrix and $(,)$ is the Euclidian inner product.

Formulating the equivalent optimal control problem associated with the estimation problem leads to the canonic equations

$$\dot{x} = g(t,x) - \frac{1}{2} k_2^{-1} \lambda \qquad (10.44)$$

$$\dot{\lambda} = 2h_2^T k_1 \{y(t) - h(t,x)\} - g_x^T \lambda \qquad (10.45)$$

where

$$h_x \triangleq \begin{bmatrix} \frac{\partial h_1}{\partial x_1} & \cdots & \frac{\partial h_1}{\partial x_n} \\ \cdot & & \\ \cdot & & \\ \frac{\partial h_m}{\partial x_1} & \cdots & \frac{\partial h_m}{\partial x_n} \end{bmatrix} \tag{10.46}$$

and

$$g_x \triangleq \begin{bmatrix} \frac{\partial g_1}{\partial x_1} & \cdots & \frac{\partial g_1}{\partial x_n} \\ \cdot & & \\ \cdot & & \\ \frac{\partial g_n}{\partial x_1} & \cdots & \frac{\partial g_n}{\partial x_n} \end{bmatrix} \tag{10.47}$$

and the superscript T denotes the transpose of a matrix. The associated boundary conditions are given by

$$\begin{aligned} \lambda(0) &= 2k_3(x(0) - m_0) \\ \lambda(T) &= 0 \ . \end{aligned} \tag{10.48}$$

Identifying the right-hand sides of equations (10.44) and (10.45) with the vector valued functions $\alpha(t,x,\lambda)$ and $\beta(t,x,\lambda)$, equations (10.44) and (10.45) can be rewritten as

$$\dot{x} = \alpha(t,x,\lambda) \tag{10.49}$$

$$\dot{\lambda} = \beta(t,x,\lambda) \quad . \tag{10.50}$$

Consider now the boundary conditions

$$\begin{aligned} \lambda(0) &= 2k_3(x(0) - m_0), \\ \lambda(T) &= c. \end{aligned} \tag{10.51}$$

Denote the value of the solution $x(t)$ at the instant t_1 by the vector valued function $\phi(t_1,T,c)$. Then it can be shown that ϕ satisfies the partial differential equation

$$\phi_T(t_1,T,c) + \phi_c^T(t_1,T,c)\beta(T,\psi(T,c),c) = 0 \tag{10.52}$$

when

$$\psi(T,c) \triangleq \phi(T,T,c) \quad . \tag{10.53}$$

The vector valued function $\psi(T,c)$ satisfies the partial differential equation

$$\begin{aligned} \psi_T(T,c) + \psi_c^T(T,c)\beta(T,\psi(T,c),c) \\ = \alpha(T,\psi(T,c),c) \end{aligned} \tag{10.54}$$

and the initial condition

$$\psi(0,c) = \frac{1}{2} k_3^{-1} c + m_0 \quad . \tag{10.55}$$

In equation (10.52)

$$\phi_c \triangleq \begin{bmatrix} \frac{\partial\phi_1}{\partial c_1} & \cdots & \frac{\partial\phi_1}{\partial c_n} \\ \cdot & & \\ \cdot & & \\ \frac{\partial\phi_n}{\partial c_1} & \cdots & \frac{\partial\phi_n}{\partial c_n} \end{bmatrix}$$

The matrix ψ_c in equation (10.54) is similarly defined.

In particular, if g and h in equations (10.42a) and (10.42b) are linear in x and take the form

$$\begin{aligned} g(t,x) &= g(t)x + f(t) \\ h(t,x) &= h(t)x \quad , \end{aligned} \tag{10.56}$$

where $g(t)$ is a $n \times n$ matrix, $f(t)$ is a n vector, and $h(t)$ is a $m \times n$ matrix, comparison of equations (10.44) and (10.45) with equations (10.49) and (10.50) yields

$$\begin{aligned} \alpha(t,x,\lambda) &= g(t)x + f(t) - \frac{1}{2}k_2^{-1}\lambda \\ \beta(t,x,\lambda) &= 2h^T(t)k_1\{y(t) - h(t)x\} - g^T(t)\lambda \ . \end{aligned} \tag{10.57}$$

Assuming a solution to equation (10.54) of the form

$$\psi(T,c) = P(T)c + q(T) \tag{10.58}$$

where $P(T)$ is a $n \times n$ matrix and $q(T)$ is a n vector leads to the differential equations

$$\dot{P}(T) = g(T)P + Pg^T(T) + 2Ph^T(T)k_1 h(T)P - \frac{1}{2} k_2^{-1} \quad (10.59)$$

$$\dot{q}(T) = g(T)q(T) + f(T) - 2P(T)h^T(T)k_1\{y(T) - h(T)q(T)\} \quad (10.60)$$

and initial conditions

$$P(0) = \frac{1}{2} k_3^{-1} \quad (10.61)$$

$$q(0) = m_0 \ . \quad (10.62)$$

Similarly, assuming a solution to equation (10.52) of the form

$$\phi(t_1,T,c) = R(t_1,T)c + r(t_1,T) \quad (10.63)$$

leads to the differential equations

$$R_T(t_1,T) = R(t_1,T)g^T(T) + 2R(t_1,T)h^T(T)k_1 h(T)P(T) \quad (10.64)$$

$$r_T(t_1,T) = -2R(t_1,T)h^T(t)k_1\{y(T) - h(T)q(T)\} \quad (10.65)$$

with the initial conditions

$$\begin{aligned} R(t_1,t_1) &= p(t_1) \\ r(t_1,t_1) &= q(t_1) \ . \end{aligned} \quad (10.66)$$

The approximate equations for the filter when g and h in equations (10.42a) and (10.42b) are not linear in x are given by assuming an approximate solution to equation (10.54) of the form specified by equation (10.58). The quantities $P(T)$ and $q(T)$ of equation (10.58) satisfy approximately the differential equations

$$\frac{dq}{dT} = g(T,q(T)) - 2P(T)h_x^T(T,q(T))k_1[y(T) - h(T,q(T))] \tag{10.67}$$

$$\frac{dP}{dT} = g_x(T,q(T))P + Pg_x^T(T,q(T)) + 2P[h_x^T(T,q(T))k_1\{y(T) - h(T,q(T))\}]_x P - \frac{1}{2}k_3^{-1} \tag{10.68}$$

and initial conditions

$$P(0) = \frac{1}{2}k_3^{-1}, \quad q(0) = m_0 . \tag{10.69}$$

Similarly assuming an approximate solution to equation (10.52) of the form given by equation (10.63) leads to a differential equations approximately satisfied by $R(t_1,T)$ and $r(t_1,T)$

$$r_T(t_1,T) = -2R(t_1,T)h_x^T(T,q(T))k_1[y(T) - h(T,q(T))] \tag{10.70}$$

$$\begin{aligned} R_T(t_1,T) &= -2R(t_1,T)[h_x^{\,T}(T,q)(T))k_1\{y(T) \\ &\quad - h(T,q(T))\}]_x P(T) + R(t_1,T)g_x^{\,T}(T,q(T)) \end{aligned} \tag{10.71}$$

with initial conditions

$$\begin{aligned} R(t_1,t_1) &= P(t_1) \\ r(t_1,t_1) &= q(t_1)\,. \end{aligned} \tag{10.72}$$

In equations (10.68) and (10.71) $[h_x^{\,T}(T,q(T))k_1\{y(T) - h(T,q(T))\}]_x$ is an $n \times n$ matrix with i^{th} column $\frac{\partial}{\partial x_i}[h_x^{\,T}(t,x)k_1 \cdot \{y(T) - h(t,x)\}]$.

REFERENCES

1. Kalman, R., and Bucy, R.S., "New Results in Linear Filtering and Prediction Theory," Journal of Basic Engineering, Trans. ASME, Series D,Vol. 83, No. 1, March 1961, pp. 96-108.

2. Bellman, R. E., Kagiwada, H. H. Kalaba, R.E., and Sridhar, R., "Invariant Imbedding and Nonlinear Filtering Theory," Journal of Astronaut Science, Vol. 13. No. 3, May 1966, pp. 110-115.

3. Detchmendy, D. M., and Sridhar, R., "Sequential Estimation of States and Parameters in Noisy Nonlinear Dynamical Systems," Journal of Basic Engineering, Trans. ASME, Series D, Vol. 88, No. 2, June 1966, pp. 362-368.

4. Pontryagin, L.S., Boltyanskii, V.G., Gamkrelidze, R.V., and Mischenko, E.F., The Mathematical Theory of Optimal Processes, Interscience Publishers, New York, 1962.

5. Bellman, R., Kagiwada, H., and Kalaba, R., "Invariant Imbedding and the Numerical Integration of Boundary-Value Problems for Unstable Linear Systems of Ordinary Differential Equations," Comm. ACM, Vol. 10, No. 2, Feb. 1967, pp. 100-102.

6. Bellman, R., Kagiwada, H., Kalaba, R., and Ueno, S., "Invariant Imbedding and the Computation of Internal Fields for Transport Processes," Journal of Mathematical Analysis and Application, Vol. 12, 1965, pp. 541-548.

7. Bellman, R., and Kalaba, R., "On the Principle of Invariant Imbedding and Propagation through Inhomogeneous Media," Proceedings of the National Academy of Sciences, U.S., Vol. 42, 1956, pp. 629-632.

Chapter 11

DETERMINATION OF TIME LAGS

11-1 INTRODUCTION

Various studies in control theory and mathematical biology [1] have led to differential-difference equations with variable time lags. A basic problem has been estimating the time lag on the basis of observations of a process that the system undergoes. Since systems of differential-difference equations with both variable and constant time lags can be reduced to systems of ordinary differential equations [2], we felt an assault on this problem was possible.

In this chapter we examine an inverse problem: Given a system of differential-difference equations and some knowledge of its solution, estimate the time lags in the equations. For simplicity, consider a single equation

$$\dot{u}(t) = g[u(t - a), t], \quad 0 \leqq t \leqq T \tag{11.1}$$

with a constant time lag a. Assume that $T = Na$, where N is an integer. We are given

$$u(t) = h(t), \quad t \leqq 0, \tag{11.2}$$

and

$$u(t_i) \overset{\sim}{\sim} \beta_i, \quad i = 1, 2, \ldots, K, \; 0 \leqq t_i \leqq T, \tag{11.3}$$

where $\beta_1, \beta_2, \ldots,$ are observations of the solution. We seek the estimate of the lag that minimizes the sum of the squares of the deviations between the observations and the theoretical values

$$\min_a S, \tag{11.4}$$

where

$$S = \sum_{i=1}^{K} [(u(t_i) - \beta_i]^2 . \tag{11.5}$$

This inverse problem will first be formulated as a nonlinear multipoint boundary-value problem for a system of ordinary differential equations. Then a method of solution utilizing quasilinearization will be outlined. Finally, these procedures will be illustrated with a simple numerical example.

11-2 FORMULATION

The differential-difference eq. (11.1) may be replaced by the system of ordinary differential equations

$$\dot{u}_1(t) = g[h(t-a), t], \quad u_1(0) = h(0), \tag{11.6}$$

$$\left.\begin{aligned} \dot{u}_n(t) &= g[u_{n-1}(t), (n-1)a+t] \\ u_n(0) &= u_{n-1}(a) \end{aligned}\right\}, \ n = 2, 3, \ldots, N, \tag{11.7}$$

for the independent variable t in the interval, $0 \leq t \leq a$. Here we view $u_n(t)$ as the solution on the nth interval of length a.

Since a is an unknown length, the interval of integration is normalized to unity by introducing a new independent variable s,

$$t = as, \quad 0 \leq s \leq 1. \tag{11.8}$$

Then we introduce new dependent variables $v_n(s)$,

$$u_n(t) = u_n(as) = v_n(s), \quad n = 1, 2, \ldots, N. \tag{11.9}$$

These new functions satisfy the system of equations

$$\dot{v}_1(s) = ag[h(as-a), as], \ v_1(0) = h(0), \tag{11.10}$$

$$\dot{v}_n(s) = ag[v_{n-1}(s), (n-1)a + as],$$

$$v_n(0) = v_{n-1}(1), \ n = 2, 3, \ldots, N. \tag{10.11}$$

Here the dot indicates differentiation with respect to s.

Note that the original observation time t_i, $0 \leq t_i \leq T$, is now related to the new variable s_i through the formula

$$t_i = (n_i - 1)a + as_i, \quad 0 \leq t_i \leq T, \tag{11.12}$$

where $(n_i - 1)$ is the largest integer for which $(n_i - 1)\, a \leq t_i$. The reader must keep in mind that a is unknown and that at each stage it is approximated by a^o: see below. In place of $u(t_i)$ we may write $v_{n_i}(s_i)$, in view of Eqs. (11.9) and (11.12). The quantity S to be minimized becomes

$$S = \sum_{i=1}^{K} [v_{n_i}(s_i) - \beta_i]^2. \tag{11.13}$$

The nonlinear multipoint boundary-value problem that we want to solve is defined by the system of Eqs. (11.10) through (11.11), the expression in Eq. (11.13), and the condition

$$\min_a S. \tag{11.14}$$

11-3 METHOD OF SOLUTION

The basic method of solution is that of quasilinearization. Quasilinearization is a successive approximation procedure in which a sequence of related linear multipoint boundary-value problems is solved. For this, a current approximation of the solution and time lag is needed. Let these be denoted by $v_n^o(s)$, $n = 1, 2, \ldots, N$, and a^o, respectively. Let the new approximation be denoted by $v_n^1(s)$, $n = 1, 2, \ldots, N$, and let the new estimate of the lag be a^1. The functions $v_n^1(s)$, $n = 1, 2, \ldots, N$ satisfy the system of linear differential equations

$$\dot{v}_1^1(s) = a^o g[h(a^o s - a^o),\ a^o s] + (a^1 - a^o)\frac{\partial}{\partial a^o} a^o\ g[h(a^o s - a^o),\ a^o s], \tag{11.15}$$

$$v_1^1(0) = h(0)\ ,$$

$$\dot{v}_n^1(s) = a^o g[v_{n-1}^o(s),\ \ (n-1)a^o + a^o s]$$

$$+ (a^1 - a^o)\frac{\partial}{\partial a^o}\ a^o\ g[v_{n-1}^o(s),\ \ (n-1)a^o + a^o s]$$

$$+ [v_{n-1}^1(s) - v_{n-1}^o(s)]\frac{\partial}{\partial v_{n-1}^o}\ a^o g[v_{n-1}^o(s),$$

$$(n-1)a^o + a^o s], \tag{11.16}$$

$$v_n^1(0) = v_{n-1}^1(1),\ \ n=2,\ 3,\ \ldots,\ N,$$

for $0 \leq s \leq 1$.

We represent the solution of Eqs. (11.15) and (11.16) as the linear combination of two particular solutions, $p_n(s)$ and $q_n(s)$,

$$v_n^1(s) = p_n(s) + a^1 q_n(s), \tag{11.17}$$

in which the new estimate of the lag appears as a multiplier of $q_n(s)$. These particular solutions are produced by numerical integration, starting with suitably chosen initial conditions. The minimization of the expression on the right-hand side of Eq. (11.13), and with Eq. (11.17) providing a substitute for $v_{n_i}(s_i)$, leads to the determination of a^1 in terms of the known quantities $p_n(s)$, $q_n(s)$, and β_i. Then the use of Eq. (11.17) produces the new approximation of the solution of the differential-difference equation.

The cycle may be repeated for higher approximations of the time lag and the solution of the lag equation.

11-4 NUMERICAL EXAMPLE

Consider the nonlinear differential-difference equation

$$\dot{u}(t) = - u^2(t-a) + \cos t + \sin^2 t, \quad t \geq 0 , \qquad (11.18)$$

with

$$u(t) = \sin t, \quad t \leq 0. \qquad (11.19)$$

The constant a is the unknown lag. Let its true value be $\pi = 3.14159265$. Then the solution of Eqs. (11.18) and (11.19) is

$$u(t) = \sin t, \quad \text{all } t, \qquad (11.20)$$

as can be readily verified.

In the inverse problem we seek to solve, Eqs. (11.18) and (11.19) and the accurate observations

$$u(t_i) \approx \beta_i, \quad i = 1, 2, \ldots, 6, \qquad (11.21)$$

are provided for $t_1 = 0$, $t_2 = \pi/2$, $t_3 = 2\pi/2$, ..., $5\pi/2$. The aim is to estimate the value of the lag a in such a way that the quantity

$$S = \sum_{i=1}^{6} (u(t_i) - \beta_i)^2 \tag{11.22}$$

is minimized.

The method of solution is that described in the preceding section. We performed a series of experiments in which we started with different estimates of a, called a^o, and with initial approximations of $u(t)$ produced by the integration of the exact differential Eqs. (11.10) and (11.11) with $a = a^o$. The starting values of a and the final values after at most six iterations are presented in Table 11-1.

Table 11-1

Numerical Results

Run	Initial estimate a^o	Final estimate of a
1	$0.9\ \pi \cong 2.83$	3.14158
2	$1.1\ \pi \cong 3.46$	3.14158
3	$1.2\ \pi \cong 3.77$	3.14158
4	$1.25\pi \cong 3.93$	Diverged

In the successful trials, Runs 1, 2, and 3, the value of the lag a correct to one part in 300,000 is obtained. In the unsuccessful trial the estimates diverge and no solution is obtained. Such trials performed on an IBM 7044 consume about $1\frac{1}{2}$ min. of computing time with $N = 5$ periods and an integration step size of $\Delta s = 0.005$ for a total of one thousand steps over an interval of length two.

11-5 DISCUSSION

The proposed method for the estimation of time lags may be extended to the case of variable lags. Lags involving the dependent variable itself may also be handled by using a successive approximation scheme.

REFERENCES

1. Bellman, R.E., J. Jacquez, and R. E. Kalaba, A Mathematical Model of Drug Distribution in the Body: Implications for Cancer Chemotherapy, Proc. 3rd Internat. Congr. Chemotherapy, pp. 1694-1707, Georg Thieme, Stuggart, 1964.

2. Bellman, R. E., J. D. Buell, and R. E. Kalaba, Numerical Integration of a Differential-Difference Equation with a Decreasing Time-lag, Commun. ACM 8(4) (1965), 227-228.

Chapter 12

IDENTIFICATION PROBLEM OF CARDIOLOGY

12-1 INTRODUCTION

Recently it has been possible to account for the potentials which are observed on the skin of a patient and which are due to ventricular depolarization [1]. The basic idea is to divide the ventricles into a number of segments, each of which contains a dipole source of current. The dipole is located in the center of the segment, and its axis is orthogonal to the surface of its segment. The dipoles have time-varying moments which are dependent upon the depolarization wave which sweeps over the ventricles. These concepts are illustrated in Figures 12-1 and 12-2. The heart is shown divided into twenty segments in Fig 12-1a. The postulated sequence of depolarization events (Fig. 12-1b) is obtained via an extrapolation of the well-known experimental results of A. Scher [2] for canine hearts. The body is viewed as a volume conductor. In this fashion surface potential time histories can be produced, from dipole moment functions such as that shown in Fig. 12-2. These surface potentials are in

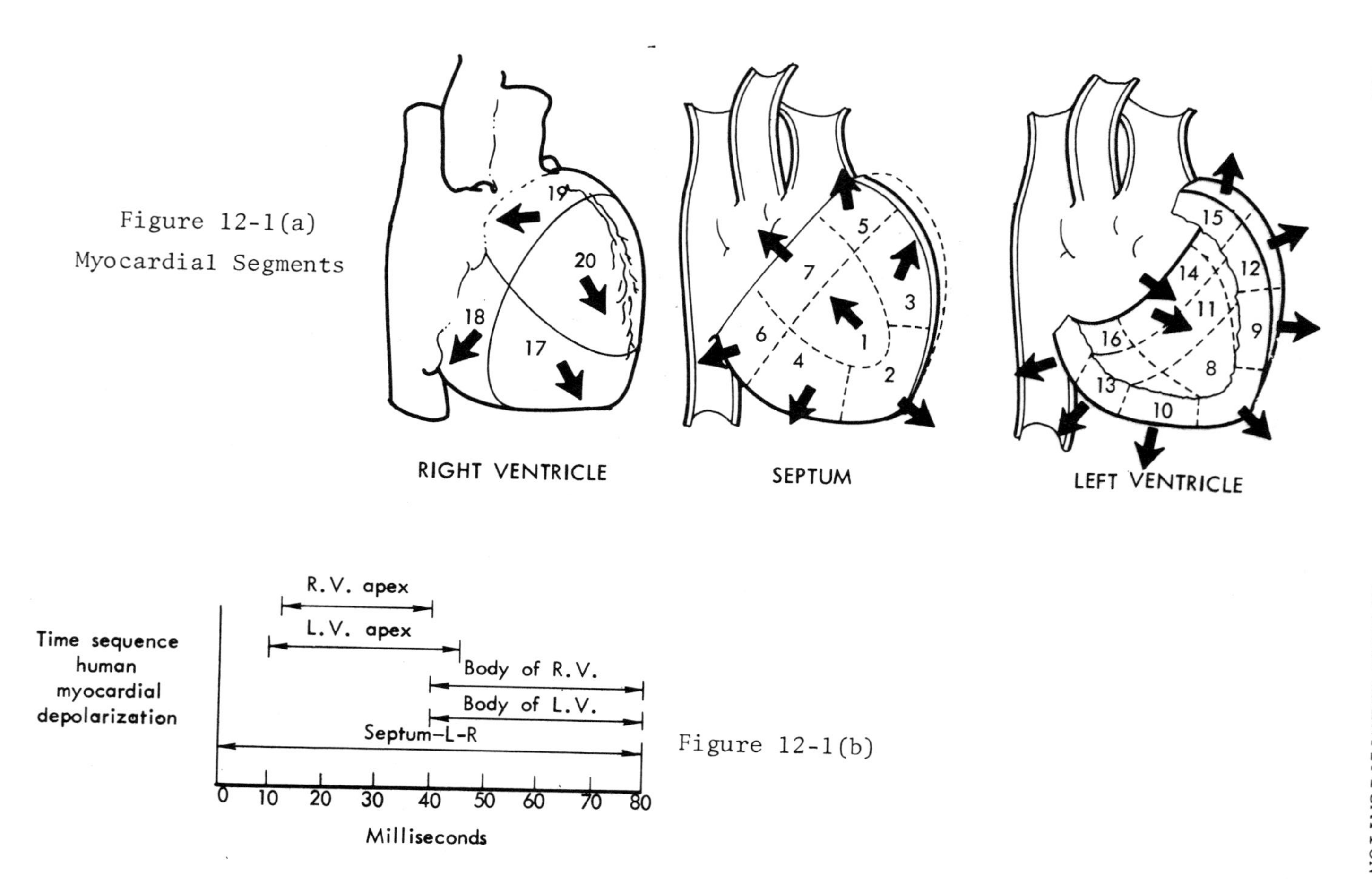

Figure 12-1(a)
Myocardial Segments

Figure 12-1(b)

good agreement with those obtained from both normal and abnormal hearts, as we have demonstrated by numerous high-speed computer runs and comparison with electro- and vectorcardiograms.

The aim of the present work is to show that it is also possible to solve the inverse problem; i.e., given the time histories of the surface potentials we can deduce the time histories of the individual dipole moments. From the mathematical viewpoint we treat this as a nonlinear multipoint boundary value problem. It is resolved using quasilinearization [3, 4] coupled with modern electronic computers.

12-2 BASIC ASSUMPTIONS

Introduce standard orthogonal reference axes, x, y, z, and consider the dipole located in the ith segment at point (x_i, y_i, z_i) and having (l_i, m_i, n_i) as the direction cosines of its axis. This produces a potential ω at the point (a_j, b_j, c_j) which depends upon the assumptions made about the body and its surrounding medium as a volume conductor. If, for simplicity, we consider the body to be homogeneous and of infinite extent, then we may write

$$w(a_j, b_j, c_j, y) = (KF_i(t)\cos\theta_{ij})r_{ij}^{-2}, \qquad (12.1)$$

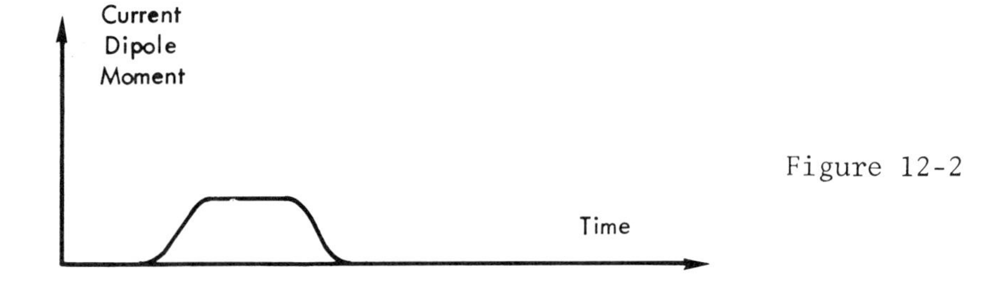

Figure 12-2

$$K = \text{const.}$$
$$F_i(t) = \text{moment of dipole in ith segment at time } t,$$
$$\cos\theta_{ij} = [l_i(a_j - x_i) + m_i(b_j - y_i) + n_i(c_j - z_i)]r_{ij}^{-1}$$
$$r_{ij}^{-2} = (a_j - x_i)^2 + (b_j - y_i)^2 + (c_j - z_i)^2.$$

Points infinitely far from the dipoles are at zero potential. Let us introduce a reference point (a_o, b_o, c_o) and assume that at N surface points the potentials with respect to that at the reference point are determined. In addition assume that the ventricles are divided into M segments.

Now we must consider the dipole moment functions $F_i(t)$, $i = 1, 2, \ldots, M$, in more detail. One is pictured in Figure 12-2. We have found that these curves are well represented as solutions of the ordinary differential equations

$$\left.\begin{aligned} dF_i/dt &= k_i(t_i - t)F_i \\ F_i(0) &= h_i \end{aligned}\right\} \begin{cases} 0 \leqq t \leqq 80 \text{ msec}, \\ i = 1, 2, \ldots, M. \end{cases} \tag{12.2}$$

The parameters t_i are the times at which the maxima of the moments occur, k_i are related to the broadness of the curves, and h_i are related to the maxima of the moments. The analytical solution of these equations is of no interest to us, since in more refined studies equations (12.2) will be replaced by much more complex differential equations.

We denote the potential produced at time t at the jth observation point by $V_j(t)$, $j = 1, 2, \ldots, N$, and $0 \leqq t \leqq 80$ msec. We have

$$V_j(t) = K \sum_{i=1}^{M} (F_i(t) \cos \theta_{ij}) r_{ij}^{-2}$$

$$- K \sum_{i=1}^{M} (F_i(t) \cos \theta_i) r_i^{-2} \qquad (12.3)$$

$$(j = 1, 2, \ldots, N).$$

where

$$\cos \theta_i$$

$$= [l_i(a_o - x_i) + m_i(b_o - y_i) + n_i(c_o - z_i)] r_i^{-1} \qquad (12.4a)$$

$$r_i^2 = (a_o - x_i)^2 + (b_o - y_i)^2 + (c_o - z_i)^2 \quad (i = 1, 2, \ldots, M)$$

$$(12.4b)$$

12-3 THE INVERSE PROBLEM

Assume that a heart has been observed for $0 \leq t \leq 80$ msec and that the surface potential time series are recorded; i.e., $b_j(t)$ = observed relative surface potential at time t at the jth observation position, $j = 1, 2, \ldots, N$ and $t = 0, 1, 2, \ldots, 80$ msec. We wish to determine the 3M constants k_i, t_i, and h_i in the differential equations (12.2) which are such that the sum of the squares of the differences of the potentials produced at the observation points and the observed potentials is as small as possible. If

$$S = \sum_{r=0}^{80} \sum_{j=1}^{N} (V_j(r) - b_j(r))^2, \qquad (12.5)$$

where the functions $V_j(t)$ are determined from equation (12.3) and $F_i(t)$ are determined from equation (12.2), we wish to minimize S through an appropriate choice of the heart parameters t_i, k_i and h_i, $i = 1, 2, \ldots, M$.

12-4 QUASILINEARIZATION

The parameters t_i and k_i appear as coefficients in equation (12.2) and the parameters h_i appear as initial conditions. We make the preliminary simplification of considering t_i and k_i to be functions of time, but subject to the conditions

$$dt_i/dt = 0 \tag{12.6}$$

$$dk_i/dt = 0 \qquad (i = 1, 2, \ldots, M). \tag{12.7}$$

We adjoin equations (12.6) and (12.7) to equation (12.2) and wish to determine initial conditions for them which result in the smallest value of S. This brings us to a problem which is abstractly equivalent to the orbit determination problem considered in [5], as well as those problems discussed in earlier chapters.

Such problems may be resolved computationally by the technique of quasilinearization. This is a successive approximation technique which requires us to solve a linear boundary value problem at each stage, and which is quadratically convergent. The linear boundary value problem is resolved computationally by producing a particular solution and a set of independent homogeneous solutions so as to minimize the sum S. The quadratic convergence implies that the number of correct digits in the approximation is asymptotically doubled from stage to stage. Details are available in [3, 4, 6, 7].

12-5 NUMERICAL EXPERIMENTS

Earlier computational work led us to believe that we could handle the inverse problem in the quasilinearization manner. Some numerical experiments were conducted to verify this. First we chose M, the number of ventricular segments, equal to five and N, the number of observation points, equal to three. Then various values were assigned to the parameters t_i, k_i and h_i, $i = 1, 2, \ldots, 5$ to simulate a normal heart and abnormal hearts exhibiting hypertrophies and infractions. Also values were assigned to (x_i, y_i, z_i), (a_j, b_j, c_j) and (l_i, m_i, n_i) using anatomical models. By integrating equations (12.2) with these values, and using (12.3) and (12.4), we produced sets of hypothetical surface potentials corresponding to a variety of hearts. We used segments 1, 8, 11, 14 and 20, and these five segments of the heart produced good simulation of potentials measured by three orthogonal leads in normal human subjects (see Fig. 12-3).

We assumed that the values t_i, $i = 1, 2, \ldots, 5$ are known, which we take to represent a heart with a normal conduction system. We also assumed a set of surface potential measurements $b_j(r)$, $j = 1, 2, 3$; $r = 0, 1, 2, \ldots, 80$. Then use of the method sketched in Section 4 did produce the "missing" values of the heart parameters, k_i and h_i, $i = 1, 2, \ldots, 5$. This is discussed in detail in [8, 9].

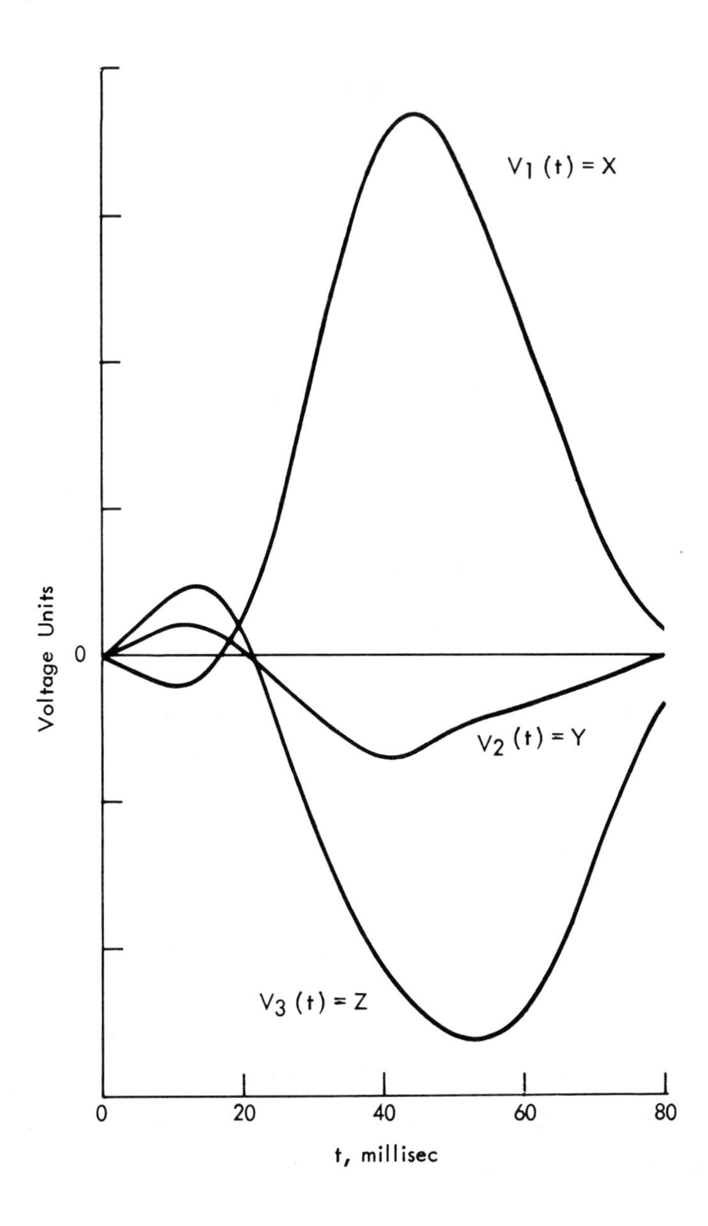

Figure 12-3 Potentials Produced by Five-Segment Normal Heart Model

A FORTRAN IV program for the IBM 7044 was written and execution required about one minute in each case. It should be borne in mind that the computational load is considerable. We were solving 110 simultaneous linear differential equations and solving linear algebraic equations of order ten at each stage in the calculation. Much remains to be done to investigate the range of convergence of the method, to determine the effects of errors in the observed surface potentials, and to make the computational program as efficient as possible. These preliminary experiments merely indicate the feasibility of this approach to the formulation and numerical solution of the inverse problem.

12-6 DISCUSSION

Our goal has been the production of a method for converting surface potential measurements into estimates of heart parameters. The feasibility of viewing this as a nonlinear multipoint boundary value problem and using quasilinearization and computing machines to effect the solution has been demonstrated. Intra-myocardial electrode techniques at open heart surgery offer interesting possibilities for experimental verification of presurgery estimates based on our models and methods.

The possible application of this approach to the computer diagnosis of clinical electrocardiograms and vectocardiograms is readily apparent. The physiological foundation of this heart model would appear to offer a more rational basis for such efforts than the currently used programs for machine diagnosis that are based on statistical and empirical data. Determining these electrical parameters that represent the state of local segments of the heart suggests applications of this technique to many public health and preventive medicine problem in heart disease. For example, the current technique suggests that the loss of 1/100 or less of the total cardiac mass may be detectable. Hence, this may be a sensitive tool for following the natural history of destructive myocardial lesions such as produced by coronary atherosclerosis and for evaluating regimens purporting to change the course of such processes, e.g., smoking, anticoagulants, exercise and low fat diets. The need for further exploration is clearly indicated.

REFERENCES

1. Selvester, R., Collier, C., and Pearson, R., Analogue Computer Model of the Vectorcardiogram. To appear.

2. Scher, A., and Yound, A., The Pathway of Ventricular Depolarization in the Dog. Circul. Res., 4 (1946), 461.

3. Bellman, R., and Kalaba, R., Quasilinearization and Boundary Value Problems, Amer. Elsevier Publ. Co. New York, 1965.

4. Kalaba, R., Some Aspects of Quasilinearization. In J. P. LaSalle and S. Lefschetz (Eds.), Nonlinear Differential Equations and Nonlinear Mechanics, Academic Press, New York, 1963.

5. Bellman, R., Kagiwada, H., and Kalaba, R., Orbit Determination as a Multi-point Boundary Value Problem and Quasilinearization. Proc. Nat. Acad. Sci. USA, 48 (1962), pp. 1327-1329.

6. Kalaba, R., On Nonlinear Differential Equations, The Maximum Operation and Monotone Convergence. J. Math. Mech., 8 (1959), pp. 519-574.

7. Bellman, R., Kagiwada, H., and Kalaba, R., A Computational Procedure for Optimal System Design and Utilization, Proc. Nat. Acad. Sci. USA, 48 (1962), pp. 1524-1528.

8. Selvester, R., R. Kalaba, C. Collier, R. Bellman and H. Kagiwada, "Simulated Myocardial Infarction with a Mathematical Model of the Heart Containing Distance and Boundary Effects," Vectorcardiography (Proc. Long Island Jewish Hospital), North Holland, Amsterdam, (1966), pp. 403-410.

9. Bellman, R., C. Collier, H. Kagiwada, R. Kalaba and R. Selvester, "A Digital Computer Model of the Vectorcardiogram with Distance and Boundary Effects - Simulated Myocardial Infarction," American Heart Journal, Vol. 74, No. 6, pp. 792-808 (1967).

APPENDIX A

PROBLEMS

Chapter 1

1. Show that, with the particular and homogeneous solutions $p(t)$ and $h^i(t)$ satisfying their prescribed initial value problems, the vector function $x(t)$ does satisfy the system of linearized differential equations and initial conditions.
2. Summarize all the equations needed in the numerical solution of the problem.
3. Write your own program to determine the unknown mass and orbit, and reproduce the results presented in Chapter 1. Try other initial approximations, as well as other sets of observations.
4. How would one select an optimal set of observations?

Chapter 2

1. Look up discussions of quadrature in the cited references or other good books on numerical methods such as

C. Lanczos, Applied Analysis, Prentice Hall, 1956.

Chapter 3

1. Determine the structure of the medium using different types of errors in observations and different criteria for fitting. Discuss the results.

Chapter 5

1. Do the numerical studies for the neutron transport problem.

Chapter 6

1. Look up the numerical inversion of Laplace transforms.
2. Solve an inverse problem for the heat equation.

$$u_t = a^2 u_{xx}.$$

Chapter 7

1. Assume the form

$$u(x,t) = \sum_{n=1}^{N} a_n(x) \sin nt,$$

and solve the inverse problem, i.e., determine the functions $a_n(x)$.

2. Remove the spatial derivative in the wave equation by use of the approximation

$$(u_n)_{xx} \cong \frac{u_{n+1}(t) - 2u_n(t) + u_{n-1}(t)}{\Delta^2}$$

and solve.

Chapter 8

1. Discuss and compare the various methods for finding the optimal set of constants (search procedure, gradient method, Newton's method, ...).

Chapter 9

1. Extend results to the vector case.

Chapter 10

1. Can you give another derivation of the basic estimator equations?

Chapter 11

1. Show that the differential-difference Eq. (11.1) may be replaced by the system of ordinary differential equations (11.6) and (11.7).

Chapter 12

1. Find a better approximation to the dipole moment function $F_i(t)$ involving as few constants as possible.

APPENDIX B

FORTRAN PROGRAMS

B-1 Remarks

This Appendix contains listings of the FORTRAN Programs used to solve the orbit determination problem of Chapter 1. and the layered medium problem of Chapter 2. They are complete but do require the addition of the library routines which are described in Appendix C.

B-2 Programs for Orbit Determination

Program A.1. Production of Observations. The complete program consists of a MAIN program and a DAUX subroutine, and requires library routine INTS/INTM.

```
$JOB            2890,3BODY,K0160,5,100,100,C
$IBJOB          MAP
$IBFTC MAIN     REF
      COMMON ALPHA,X1,Y1,C(4),T(51)
C
C             3 BODY ORBIT DETERMINATION
C
    1 READ(5,100)NPRNT,MPRNT,ALPHA,X1,Y1,DELTA
      WRITE(6,90)NPRNT,MPRNT,ALPHA,X1,Y1,DELTA
      READ(5,101)(C(I),I=1,4)
      WRITE(6,91)(C(I),I=1,4)
C
      T(2)=0.0
      T(3)=DELTA
      DO 2 L=4,7
    2 T(L)=C(L-3)
      CALL INTS(T,4,2,0,0,0,0,0,0)
      THETA=ATAN2(T(6),T(4)-1.0)
      SN=SIN(THETA)
      CS=COS(THETA)
      TN=SN/CS
      WRITE(6,92)
      WRITE(6,93)T(2),T(4),T(5),T(6),T(7),THETA,TN
C
      DO 4 M1=1,MPRNT
      DO 3 M2=1,NPRNT
    3 CALL INTM
      THETA=ATAN2(T(6),T(4)-1.0)
      SN=SIN(THETA)
      CS=COS(THETA)
      TN=SN/CS
    4 WRITE(6,93)T(2),T(4),T(5),T(6),T(7),THETA,TN
      GO TO 1
C
  100 FORMAT(2I12,4E12.8)
  101 FORMAT(6E12.8)
   90 FORMAT(1H12I20,4E20.8)
   91 FORMAT(1H06E20.8)
   92 FORMAT(///9X1HT,19X1HX,15X5HDX/DT,19X1HY,15X5HDY/DT,15X5HTHETA
     1,13X7HTANGENT//)
   93 FORMAT(F10.2,1P6E20.5)
      END
$IBFTC DAUX     REF
      SUBROUTINE DAUX
      COMMON ALPHA,X1,Y1,C(4),T(51)
C
      R= T(4)**2 + T(6)**2
      R=SQRT(R**3)
      R1=(X1-T(4))**2 + (Y1-T(6))**2
      R1=SQRT(R1**3)
      T(8)=T(5)
      T(9)=-T(4)/R + ALPHA*(X1-T(4))/R1
      T(10)=T(7)
      T(11)=-T(6)/R + ALPHA*(Y1-T(6))/R1
      RETURN
      END
$ENTRY          MAIN
          10          25          0.2          4.0          1.0          0.01
         2.0         0.0          0.0          0.5
```

Program A.2. Determination of Orbit. The complete program is listed, and consists of a MAIN program, and subroutines INPUT, DAUX, FUN1, FUN2, PDR1, PDR2, and START. The library routines INTS/INTM and MATINV are required.

```
$IBFTC MAIN     REF
      COMMON T(363),NEQ,KMAX,HGRID,NGRID(5),THETA(5),W(4,251),ALPHA,
     1  H(5,5,251),P(5,251),A(50,50),B(50,1),X,U,Y,V,NPRNT,MPRNT,DTIME
      DIMENSION PIVOT(50),INDEX(50,2),IPIVOT(50)
C
C
C             THREE BODY ORBIT DETERMINATION
    1 CALL INPUT
      DO 8 I=1,5
      THET=THETA(I)
      ST=SIN(THET)
      CT=COS(THET)
      TN=ST/CT
    8 PRINT114,THET,TN
    2 CALL START
C                      K ITERATIONS
    3 DO 19 K=1,KMAX
      NEQ=30
    4 T(2)=0.0
      T(3)=HGRID
      DO 5 I=4,363
    5 T(I)=0.
      T(5)=1.0
      T(12)=1.0
      T(19)=1.0
      T(26)=1.0
      T(33)=1.0
      N=1
      X=W(1,1)
      Y=W(3,1)
    6 CALL INTS(T,NEQ,2,0,0,0,0,0,0)
        L=3
      DO 7 I=1,5
        L=L+1
      P(I,N)=T(L)
      DO 7 J=1,5
        L=L+1
    7 H(J,I,N)=T(L)
      PRINT49,T(2),((H(J,I,N),I=1,5),J=1,5)
C
C                 INTEGRATE OVER RANGE
C
      DO 11 M1=1,MPRNT
      DO 10 M2=1,NPRNT
        CALL INTM
      N=N+1
        X=W(1,N)
        Y=W(3,N)
C                 STORE P'S AND H'S
        L=3
      DO 10 I=1,5
        L=L+1
      P(I,N)=T(L)
      DO 10 J=1,5
        L=L+1
```

```
   10   H(J,I,N)=T(L)
        PRINT49,T(2),((H(J,I,N),I=1,5),J=1,5)
   11   CONTINUE
C
C                 COMPUTE CONSTANTS
        DO 14 I=1,5
          N=NGRID(I)
          THET=THETA(I)
        STHET=SIN(THET)
        CTHET=COS(THET)
        DO 13 J=1,5
   13     A(I,J)=H(J,1,N)*STHET -H(J,3,N)*CTHET
          B(I,1)=(1.-P(1,N))*STHET +P(3,N)*CTHET
   14   PRINT114,(A(I,JJ),JJ=1,5),B(I,1)
   15   CALL MATINV(A,5,B,1,DETERM,PIVOT,INDEX,IPIVOT)
        PRINT114,(B(I,1),I=1,5)
C
C                 COMPUTE NEW W'S
        N=1
        DO 20 I=1,4
   20   W(I,N)=B(I,1)
        ALPHA=B(5,1)
        PRINT40,K,ALPHA
        TIME=0.0
        AT=ATAN2(W(3,N),W(1,N) - 1.0)
        TN=W(3,N)/(W(1,N)-1.0)
        PRINT50,TIME,(W(I,1),I=1,4),AT,TN
        DO 18 M1=1,MPRNT
        DO 17 M2=1,NPRNT
        N=N+1
          DO 17 I=1,4
          W(I,N)=P(I,N)
          DO 17 J=1,5
   17     W(I,N)=W(I,N) + B(J,1)*H(J,I,N)
        TIME=TIME+DTIME
        AT=ATAN2(W(3,N),W(1,N) - 1.0)
        TN=W(3,N)/(W(1,N)-1.0)
   18   PRINT50,TIME,(W(I,N),I=1,4),AT,TN
C
   19   CONTINUE
C
          GO TO 1
   40     FORMAT(1H0/40X 9HITERATION,I3,5X7HALPHA =,    E18.6//
      1                                     6X4H    T,14X1HX,19X2HX',18X1HY,
      2                              19X2HY',15X5HANGLE,13X7HTANGENT)
   49   FORMAT(1H0F9.2,5E20.8/(10X5E20.8))
   50   FORMAT(F10.2,6E20.6)
  114   FORMAT(1H0  6E20.6)
            END
$IBFTC INPUT   REF
        SUBROUTINE INPUT
        COMMON T(363),NEQ,KMAX,HGRID,NGRID(5),THETA(5),W(4,251),ALPHA,
      1  H(5,5,251),P(5,251),A(50,50),B(50,1),X,U,Y,V,NPRNT,MPRNT,DTIME
C
        READ110,NPRNT,MPRNT,KMAX
```

```
      PRINT10,NPRNT,MPRNT,KMAX
      READ111,HGRID,ALPHA
      PRINT11,HGRID,ALPHA
      F=NPRNT
      DTIME=F*HGRID
      READ120,(NGRID(I),THETA(I),I=1,5)
      PRINT20,(NGRID(I),THETA(I),I=1,5)
 110  FORMAT(6I12)
  10  FORMAT(1H06I20)
 111  FORMAT(6E12.8)
  11  FORMAT(1H06E20.8)
 120  FORMAT(I12,E12.8)
  20  FORMAT(I20,E20.8)
        RETURN
          END
$IBFTC DAUX     REF
      SUBROUTINE DAUX
C
      COMMON T(363),NEQ,KMAX,HGRID,NGRID(5),THETA(5),W(4,251),ALPHA,
     1  H(5,5,251),P(5,251),A(50,50),B(50,1),X,U,Y,V,NPRNT,MPRNT,DTIME
     2  ,IFLAG
      DIMENSION XX(2),YY(2),ANS(2),PP(5),HH(5,5),PD(5),HD(5,5),PD1(3),
     1  PD2(3),AA(2)
C
      GO TO (10,20),IFLAG
C
  10  XX(1)=T(4)
      XX(2)=0.0
      YY(1)=T(6)
      YY(2)=0.0
      AA(1)=ALPHA
      AA(2)=0.0
      T(8)=T(5)
      CALL FUN1(XX,YY,AA,ANS)
      T(9)=ANS(1)
C
      T(10)=T(7)
      CALL FUN2(XX,YY,AA,ANS)
      T(11)=ANS(1)
      RETURN
C
  20  XX(1)=X
      XX(2)=0.0
      YY(1)=Y
      YY(2)=0.0
      AA(1)=ALPHA
      AA(2)=0.0
        L=3
      DO 1 I=1,5
        L=L+1
      PP(I)=T(L)
      DO 1 J=1,5
        L=L+1
   1  HH(J,I)=T(L)
C           DX/DT
```

```
      CALL FUN1(XX,YY,AA,ANS)
      CALL PDR1(XX,YY,AA,PD1)
      PD(1)=PP(2)
      PD(2)=ANS(1) + (PP(1)-X)*PD1(1) + (PP(3)-Y)*PD1(2)
     1  + (PP(5) - ALPHA)*PD1(3)
      DO 2 J=1,5
      HD(J,1)=HH(J,2)
    2 HD(J,2)=HH(J,1)*PD1(1) + HH(J,3)*PD1(2) + HH(J,5)*PD1(3)
C
C            DY/DT
      CALL FUN2(XX,YY,AA,ANS)
      CALL PDR2(XX,YY,AA,PD2)
      PD(3)=PP(4)
      PD(4)=ANS(1) + (PP(1)-X)*PD2(1) + (PP(3)-Y)*PD2(2)
     1  + (PP(5) - ALPHA)*PD2(3)
      DO 3 J=1,5
      HD(J,3)=HH(J,4)
    3 HD(J,4)=HH(J,1)*PD2(1) + HH(J,3)*PD2(2) + HH(J,5)*PD2(3)
C
      PD(5)=0.0
      DO 5 J=1,5
    5 HD(J,5)=0.C
C
      DO 4 I=1,5
        L=L+1
      T(L)=PD(I)
      DO 4 J=1,5
        L=L+1
    4 T(L)=HD(J,I)
      RETURN
      END
$IBFTC FUN1    REF
      SUBROUTINE FUN1(XX,YY,AA,ANS)
      DIMENSION XX(2),YY(2),AA(2),ANS(2)
      X=XX(1)
      Y=YY(1)
      A=AA(1)
      R13=(X**2 + Y**2)**1.5
      R23=((X-4.0)**2 + (Y-1.0)**2)**1.5
      ANS(1)=-X/R13 - A*(X-4.0)/R23
      RETURN
      END
$IBFTC FUN2    REF
      SUBROUTINE FUN2(XX,YY,AA,ANS)
      DIMENSION XX(2),YY(2),AA(2),ANS(2)
      X=XX(1)
      Y=YY(1)
      A=AA(1)
      R13=(X**2 + Y**2)**1.5
      R23=((X-4.0)**2 + (Y-1.0)**2)**1.5
      ANS(1)=-Y/R13 - A*(Y-1.0)/R23
      RETURN
      END
$IBFTC PDR1    REF
      SUBROUTINE PDR1(XX,YY,AA,PD1)
```

```
      DIMENSION XX(2),YY(2),AA(2),PD1(3)
      RR=(X-4.0)**2    + (Y-1.0)**2
      R25=RR**2.5
      X=XX(1)
      Y=YY(1)
      R13=RR**1.5
      A=AA(1)
      R15=RR**2.5
      RR=(X-4.0)**2    + (Y-1.0)**2
      R23=RR**1.5
      R25=RR**2.5
      PD1(1)=-1.0/R13 + 3.0*X**2/R15 - A/R23 + 3.0*A*(X-4.0)**2/R25
      PD1(2)=3.0*X*Y/R15 + 3.0*A*(X-4.0)*(Y-1.0)/R25
      PD1(3)=-(X-4.0)/R23
      RETURN
      END
$IBFTC PDR2    REF
      SUBROUTINE PDR2(XX,YY,AA,PD2)
      DIMENSION XX(2),YY(2),AA(2),PD2(3)
      X=XX(1)
      Y=YY(1)
      A=AA(1)
      RR=X**2 + Y**2
      R13=RR**1.5
      R23=RR**1.5
      RR=X**2 + Y**2
      R15=RR**2.5
      PD2(1)=3.0*X*Y/R15 + 3.0*A*(X-4.0)*(Y-1.0)/R25
      PD2(2)=-1.0/R13 + 3.0*Y**2/R15 - A/R23 + 3.0*A*(Y-1.0)**2/R25
      PD2(3)=-(Y-1.0)/R23
      RETURN
      END
$IBFTC START   REF
      SUBROUTINE START
      COMMON T(363),NEQ,KMAX,HGRID,NGRID(5),THETA(5),W(4,251),ALPHA,
     1  H(5,5,251),P(5,251),A(50,50),B(50,1),X,U,Y,V,NPRNT,MPRNT,DTIME
     2  ,IFLAG
C
      IFLAG=1
      K=0
      PRINT40,K
      N=1
      TIME=0.0
      T(2)=0.0
      T(3)=HGRID
      READ110,(T(I),I=4,7)
  110 FORMAT(6E12.8)
      CALL INTS(T,4,2,0,0,0,0,0,0)
      DO 3 I=1,4
    3 W(I,1)=T(I+3)
      PRINT50,TIME,(W(I,N),I=1,4)
      DO 2 M1=1,MPRNT
      DO 1 M2=1,NPRNT
      N=N+1
      CALL INTM
```

```
      DO 4 I=1,4
    4 W(I,N)=T(I+3)
    1 CONTINUE
      TIME=TIME+DTIME
    2 PRINT50,TIME,(W(I,N),I=1,4)
      IFLAG=2
        RETURN
C
   40   FORMAT(1H0/65X 9HITERATION,I3//
     1                                 26X4H   T,14X1HX,19X2HX',18X1HY,
     2                            19X2HY')
   50 FORMAT(F30.2,4E20.6)
          END
```

3. Programs for Identification of Layered Media.

Program B.1. Determination of c, the Thickness of the Lower Layer. The complete program, listed, consists of a MAIN program, and subroutines DAUX, NONLIN, PANDH, LINEAR, OUTPUT, and ALBEDO. The library routine INTS/INTM is required.

```
$IBFTC RTINV
      COMMON N,RT(7),WT(7),WR(7),AR(7,7),NPRNT,M1MAX,KMAX,DELTA,XTAU,
     1 ZERLAM,XLAM(2),B2(7,7),R2(7,7),IFLAG,R(28,101),T(1491),SIG,
     2 P(28,101),H(28,3,101),PTAU,PLAM(2),HTAU(3),HLAM(2,3),P2(7,7),
     3 H2(7,7,3),CONST(3),NEQ
C
C            PHASE I
C
    1 READ1000,N
      PRINT899
      PRINT900,N
      READ1001,(RT(I),I=1,N)
      PRINT901,(RT(I),I=1,N)
      READ1001,(WT(I),I=1,N)
      PRINT901,(WT(I),I=1,N)
      DO 2 I=1,N
      WR(I)=WT(I)/RT(I)
      DO 2 J=1,N
    2 AR(I,J)= 1.0/RT(I) + 1.0/RT(J)
C
  899 FORMAT(1H146X36HRADIATIVE TRANSFER - INVERSE PROBLEM / )
 1000 FORMAT(6I12)
  900 FORMAT(6I20)
 1001 FORMAT(6E12.8)
  901 FORMAT(6E20.8)
      READ1000,NPRNT,M1MAX,KMAX
      PRINT900,NPRNT,M1MAX,KMAX
      READ1001,DELTA
      PRINT901,DELTA
      READ1001,XTAU,ZERLAM,XLAM(1),XLAM(2)
      PRINT902
      PRINT903,XTAU,ZERLAM,XLAM(1),XLAM(2)
  902 FORMAT(1H123HPHASE I - TRUE SOLUTION /)
  903 FORMAT(1H0/
     1    1X11HTHICKNESS =, F10.4 /
     2    1X11HALBEDO(X) =, 20HA + B*TANH(10*(X-C)) //
     3 1X3HA =, E16.8, 10X3HB =, E16.8, 10X3HC =, E16.8 //)
      CALL NONLIN
      DO 3 I=1,N
      DO 3 J=1,N
    3 B2(I,J)=R2(I,J)
C
C
C            PHASE II
C
    4 READ1001,XTAU,ZERLAM,XLAM(1),XLAM(2)
      K=0
      PRINT904,K
      PRINT903,XTAU,ZERLAM,XLAM(1),XLAM(2)
C
      CALL NONLIN
C
  904 FORMAT(1H1   13HAPPROXIMATION, I3/ )
C
C            QUASILINEARIZATION ITERATIONS
```

```
C
      DO 5 K1=1,KMAX
      PRINT904,K1
      CALL PANDH
      CALL LINEAR
    5 CONTINUE
C
C
C
      READ1000,IGO
      GO TO (1,4),IGO
      END
$IBFTC DAUX
      SUBROUTINE DAUX
      DIMENSION V2(7,7),X(3),F(7),G(7)
      COMMON N,RT(7),WT(7),WR(7),AR(7,7),NPRNT,M1MAX,KMAX,DELTA,XTAU,
     1  ZERLAM,XLAM(2),B2(7,7),R2(7,7),IFLAG,R(28,101),T(1491),SIG,
     2  P(28,101),H(28,3,101),PTAU,PLAM(2),HTAU(3),HLAM(2,3),P2(7,7),
     3  H2(7,7,3),CONST(3),NEQ
      GO TO (1,2),IFLAG
C
CNONLINEAR
C
    1 L=3
      DO 4 I=1,N
      DO 4 J=1,I
      L=L+1
    4 V2(I,J)=T(L)
      DO 5 I=1,N
      DO 5 J=I,N
    5 V2(I,J)=V2(J,I)
      L=L+1
      VLAM2=T(L)
      SIG=T(2)
      Y=XTAU*SIG
      X(1)=ZERLAM
      X(2)=XLAM(1)
      X(3)=VLAM2
      CALL ALBEDO(Y,X,Z)
      ZLAMDA=Z
C
      DO 6 I=1,N
      F(I)=0.0
      DO 7 K=1,N
    7 F(I)=F(I) + WR(K)*V2(I,K)
    6 F(I)=0.5*F(I) + 1.0
C
      DO 8 I=1,N
      DO 8 J=1,I
      L=L+1
      DR=-AR(I,J)*V2(I,J) + ZLAMDA*F(I)*F(J)
    8 T(L)=DR
      DO 9 I=1,1
      L=L+1
    9 T(L)=0.0
```

```
      RETURN
C
C
CLINEAR
C
    2 SIG=T(2)
      Y=XTAU*SIG
      X(1)=ZERLAM
      X(2)=XLAM(1)
      X(3)=XLAM(2)
      CALL ALBEDO(Y,X,Z)
      ZLAMDA=Z
C
      DO 16 I=1,N
      F(I)=0.0
      DO 17 K=1,N
   17 F(I)=F(I) + WR(K)*R2(I,K)
   16 F(I)=0.5*F(I) + 1.0
C
CP'S
C
      L=3
      DO 14 I=1,N
      DO 14 J=1,I
      L=L+1
   14 V2(I,J)=T(L)
      DO 15 I=1,N
      DO 15 J=I,N
   15 V2(I,J)=V2(J,I)
      L=L+1
      VLAM2=T(L)
C
      DO 10 I=1,N
      G(I)=0.0
      DO 10 K=1,N
   10 G(I)=G(I) + (V2(I,K)-R2(I,K))*WR(K)
      ARG=10.0*(Y-XLAM(2))
      XTANX=-10.0*XLAM(1)*(1.0-(TANH(ARG))**2)
        M=3+NEQ
      DO 12 I=1,N
      DO 12 J=1,I
      FIJ=F(I)*F(J)
      CAPF=-AR(I,J)*R2(I,J) + ZLAMDA*FIJ
      T1=       -AR(I,J)*(V2(I,J)-R2(I,J))
      T2=       0.5*ZLAMDA*(F(I)*G(J)+F(J)*G(I))
      T3=       CAPF
      T4=(VLAM2-XLAM(2))*XTANX*FIJ
        M=M+1
   12 T(M)=T1+T2+T3+T4
      DO 19 I=1,1
        M=M+1
   19 T(M)=0.0
C
CH'S
C
```

```
      DO 100 K=1,1
C
      DO 24 I=1,N
      DO 24 J=1,I
      L=L+1
   24 V2(I,J)=T(L)
      DO 25 I=1,N
      DO 25 J=I,N
   25 V2(I,J)=V2(J,I)
      L=L+1
      VLAM2=T(L)
C
      DO 20 I=1,N
      G(I)=0.0
      DO 20 J=1,N
   20 G(I)=G(I) +  V2(I,J)*WR(J)
C
      DO 22 I=1,N
      DO 22 J=1,I
      FIJ=F(I)*F(J)
      T1=      -AR(I,J)*V2(I,J)
      T2=      0.5*ZLAMDA*(F(I)*G(J)+F(J)*G(I))
      T3=0.0
      T4=VLAM2*XTANX*FIJ
        M=M+1
   22 T(M)=T1+T2+T3+T4
C
      DO 29 I=1,1
        M=M+1
   29 T(M)=0.0
  100 CONTINUE
      RETURN
      END
$IBFTC NONLIN
      SUBROUTINE NONLIN
      COMMON N,RT(7),WT(7),WR(7),AR(7,7),NPPNT,M1MAX,KMAX,DELTA,XTAU,
     1  ZERLAM,XLAM(2),B2(7,7),R2(7,7),IFLAG,R(28,101),T(1491),SIG,
     2  P(28,101),H(28,3,101),PTAU,PLAM(2),HTAU(3),HLAM(2,3),P2(7,7),
     3  H2(7,7,3),CONST(3),NEQ
C           NONLINEAR D.E. FOR TRUE SOLUTION OR FOR INITIAL APPROX.
C
      IFLAG=1
      T(2)=0.0
      T(3)=DELTA
      M=1
      L1=0
      L3=3
      DO 1 I=1,N
      DO 1 J=1,I
      L1=L1+1
      L3=L3+1
      R2(I,J)=0.0
      R(L1,M)=R2(I,J)
    1 T(L3)=R2(I,J)
      L3=L3+1
```

```
    2   T(L3)=XLAM(2)
C
        NEQ=(N*(N+1))/2 + 1
        CALL INTS(T,NEQ,2,0,0,0,0,0,0)
C
        SIG=T(2)
        CALL OUTPUT
C
        DO 5 M1=1,M1MAX
        DO 4 M2=1,NPRNT
        CALL INTM
        M=M+1
        L1=0
        L3=3
        DO 3 I=1,N
        DO 3 J=1,I
        L1=L1+1
        L3=L3+1
        R2(I,J)=T(L3)
    3   R(L1,M)=R2(I,J)
    4   SIG=T(2)
    5   CALL OUTPUT
C
        RETURN
        END
$IBFTC PANDH
        SUBROUTINE PANDH
        COMMON N,RT(7),WT(7),WR(7),AR(7,7),NPRNT,M1MAX,KMAX,DELTA,XTAU,
       1  ZERLAM,XLAM(2),B2(7,7),R2(7,7),IFLAG,R(28,101),T(1491),SIG,
       2  P(28,101),H(28,3,101),PTAU,PLAM(2),HTAU(3),HLAM(2,3),P2(7,7),
       3  H2(7,7,3),CONST(3),NEQ
        IFLAG=2
        T(2)=0.0
        T(3)=DELTA
        M=1
C P'S
C
        L1=0
        L3=3
        DO 1 I=1,N
        DO 1 J=1,I
        L1=L1+1
        L3=L3+1
        P(L1,M)=0.0
    1   T(L3)=P(L1,M)
        L3=L3+1
        PLAM(2)=0.0
    2   T(L3)=PLAM(2)
C
C H'S
C
        DO 7 K=1,1
        L1=0
        DO 3 I=1,N
        DO 3 J=1,I
```

```
      L1=L1+1
      L3=L3+1
      H(L1,K,M)=0.0
    3 T(L3)=H(L1,K,M)
C
      L3=L3+1
    6 HLAM(2,K)=1.0
    7 T(L3)=HLAM(2,K)
C
      L=0
      DO 8 I=1,N
      DO 8 J=1,I
      L=L+1
    8 R2(I,J)=R(L,M)
      DO 9 I=1,N
      DO 9 J=I,N
    9 R2(I,J)=R2(J,I)
C
      NEQ=2*((N*(N+1))/2 + 1)
      CALL INTS(T,NEQ,2,0,0,0,0,0,0)
      LMAX=(N*(N+1))/2
      PRINT52,T(2),(P(L,M),H(L,1,M),L=1,LMAX)
   52 FORMAT(1H0F9.4,5E20.8/(10X5E20.8))
C
      DO 51 M1=1,M1MAX
      DO 50 M2=1,NPRNT
      CALL INTM
      M=M+1
CPREV.APPROX. R(I,J)
      L1=0
      DO 10 I=1,N
      DO 10 J=1,I
      L1=L1+1
   10 R2(I,J)=R(L1,M)
      DO 11 I=1,N
      DO 11 J=I,N
   11 R2(I,J)=R2(J,I)
      L1=0
      L3=3
      DO 12 I=1,N
      DO 12 J=1,I
      L1=L1+1
      L3=L3+1
   12 P(L1,M)=T(L3)
      L3=L3+1
      DO 13 K=1,1
      L1=0
      DO 14 I=1,N
      DO 14 J=1,I
      L1=L1+1
      L3=L3+1
   14 H(L1,K,M)=T(L3)
   13 L3=L3+1
   50 CONTINUE
   51 PRINT52,T(2),(P(L,M),H(L,1,M),L=1,LMAX)
```

```
      RETURN
      END
$IBFTC LINEAR
      SUBROUTINE LINEAR
      DIMENSION CHKI(3)
      DIMENSION A(49,3),B(49),EMAT(50,50),          PIVOT(50),INDEX(50,2)
     1,IPIVOT(50),FVEC(50,1)
      COMMON N,RT(7),WT(7),WR(7),AR(7,7),NPRNT,M1MAX,KMAX,DELTA,XTAU,
     1  ZERLAM,XLAM(2),B2(7,7),R2(7,7),IFLAG,R(28,101),T(1491),SIG,
     2  P(28,101),H(28,3,101),PTAU,PLAM(2),HTAU(3),HLAM(2,3),P2(7,7),
     3  H2(7,7,3),CONST(3),NEQ
CBOUNDARY CONDITIONS
      MLAST=NPRNT*M1MAX + 1
      DO 1 K=1,1
      L=0
      DO 2 I=1,N
      DO 2 J=1,I
      L=L+1
    2 H2(I,J,K)=H(L,K,MLAST)
      DO 1 I=1,N
      DO 1 J=I,N
    1 H2(I,J,K)=H2(J,I,K)
      L=0
      DO 3 I=1,N
      DO 3 J=1,I
      L=L+1
    3 P2(I,J)=P(L,MLAST)
      DO 4 I=1,N
      DO 4 J=I,N
    4 P2(I,J)=P2(J,I)
CLEAST SQUARES
      DO 5 K=1,1
      L=0
      DO 5 I=1,N
      DO 5 J=1,N
      L=L+1
    5 A(L,K)=H2(I,J,K)
      L=0
      DO 6 I=1,N
      DO 6 J=1,N
      L=L+1
    6 B(L)=B2(I,J) - P2(I,J)
C
      LMAX=N**2
      PRINT60
   60 FORMAT(1H0)
      DO 61 L=1,LMAX
   61 PRINT82,(A(L,K),K=1,1),B(L)
C
      DO 8 I=1,1
      DO 7 J=1,1
      SUM=0.0
      DO 9 L=1,LMAX
    9 SUM=SUM + A(L,I)*A(L,J)
    7 EMAT(I,J)=SUM
```

```
      SUM=0.0
      DO 10 L=1,LMAX
   10 SUM=SUM + A(L,I)*B(L)
    8 FVEC(I,1)=SUM
C
      PRINT60
      DO 81 I=1,1
   81 PRINT82,(EMAT(I,J),J=1,1),FVEC(I,1)
   82 FORMAT(10X6E20.8)
C
      FVEC(1,1)=FVEC(1,1)/EMAT(1,1)
C
      DO 11 I=1,1
   11 CONST(I)=FVEC(I,1)
C
      XLAM(2)=CONST(1)
      PRINT903,XTAU,ZERLAM,XLAM(1),XLAM(2)
  903 FORMAT(1H0/
     1     1X11HTHICKNESS =, F10.4 /
     2     1X11HALBEDO(X) =, 20HA + B*TANH(10*(X-C))   //
     3  1X3HA =, E16.8, 10X3HB =, E16.8, 10X3HC =, E16.8 //)
C
CNEW APPROXIMATION
C
      M=1
      L=0
      DO 12 I=1,N
      DO 12 J=1,I
      L=L+1
      SUM=P(L,M)
      DO 13 K=1,1
   13 SUM =SUM + CONST(K)*H(L,K,M)
   12 R(L,M)=SUM
      L=0
      DO 14 I=1,N
      DO 14 J=1,I
      L=L+1
   14 R2(I,J)=R(L,M)
      SIG=0.0
      CALL OUTPUT
C
      DO 50 M1=1,M1MAX
      DO 18 M2=1,NPRNT
      M=M+1
      L=0
      DO 15 I=1,N
      DO 15 J=1,I
      L=L+1
      SUM=P(L,M)
      DO 16 K=1,1
   16 SUM=SUM + CONST(K)*H(L,K,M)
   15 R(L,M)=SUM
      L=0
      DO 17 I=1,N
      DO 17 J=1,I
```

```
      L=L+1
   17 R2(I,J)=R(L,M)
   18 SIG=SIG + DELTA
   50 CALL OUTPUT
C
      RETURN
      END
$IBFTC OUTPUT
      SUBROUTINE OUTPUT
      DIMENSION X(3)
      COMMON N,RT(7),WT(7),WR(7),AR(7,7),NPRNT,M1MAX,KMAX,DELTA,XTAU,
     1  ZERLAM,XLAM(2),B2(7,7),R2(7,7),IFLAG,R(28,101),T(1491),SIG,
     2  P(28,101),H(28,3,101),PTAU,PLAM(2),HTAU(3),HLAM(2,3),P2(7,7),
     3  H2(7,7,3),CONST(3),NEQ
      DO 1 I=1,N
      DO 1 J=I,N
    1 R2(I,J)=R2(J,I)
      Y=XTAU*SIG
      X(1)=ZERLAM
      X(2)=XLAM(1)
      X(3)=XLAM(2)
      CALL ALBEDO(Y,X,Z)
      PRINT100, SIG,Y,Z
  100 FORMAT(1H0 7HSIGMA =,F6.2, 4X5HTAU =, F6.2, 4X8HALBEDO =,F6.2/)
      DO 2 J=1,N
    2 PRINT101,J,(R2(I,J),I=1,N)
  101 FORMAT(I10, 7F10.6)
      RETURN
      END
$IBFTC ALBEDO
      SUBROUTINE ALBEDO(Y,X,Z)
      DIMENSION X(3)
      COMMON N,RT(7),WT(7),WR(7),AR(7,7),NPRNT,M1MAX,KMAX,DELTA,XTAU,
     1  ZERLAM,XLAM(2),B2(7,7),R2(7,7),IFLAG,R(28,101),T(1491),SIG,
     2  P(28,101),H(28,3,101),PTAU,PLAM(2),HTAU(3),HLAM(2,3),P2(7,7),
     3  H2(7,7,3),CONST(3),NEQ
      ARG=10.0*(Y-X(3))
      Z=X(1) + X(2)*TANH(ARG)
      RETURN
      END
```

Program B.2. Determination of T, the Overall Optical Thickness. The complete program consists of a MAIN program and subroutines DAUX, NONLIN, PANDH, LINEAR, OUTPUT, and ALBEDO. The library routine INTS/INTM is required.

```
$IBFTC RTINV
      COMMON N,RT(7),WT(7),WR(7),AR(7,7),NPRNT,M1MAX,KMAX,DELTA,XTAU,
     1  ZERLAM,XLAM(2),B2(7,7),R2(7,7),IFLAG,R(28,101),T(1491),SIG,
     2  P(28,101),H(28,3,101),PTAU,PLAM(2),HTAU(3),HLAM(2,3),P2(7,7),
     3  H2(7,7,3),CONST(3),NEQ
C
C             PHASE I
C
    1 READ1000,N
      PRINT899
      PRINT900,N
      READ1001,(RT(I),I=1,N)
      PRINT901,(RT(I),I=1,N)
      READ1001,(WT(I),I=1,N)
      PRINT901,(WT(I),I=1,N)
      DO 2 I=1,N
      WR(I)=WT(I)/RT(I)
      DO 2 J=1,N
    2 AR(I,J)= 1.0/RT(I) + 1.0/RT(J)
C
  899 FORMAT(1H146X36HRADIATIVE TRANSFER - INVERSE PROBLEM / )
 1000 FORMAT(6I12)
  900 FORMAT(6I20)
 1001 FORMAT(6E12.8)
  901 FORMAT(6E20.8)
      READ1000,NPRNT,M1MAX,KMAX
      PRINT900,NPRNT,M1MAX,KMAX
      READ1001,DELTA
      PRINT901,DELTA
      READ1001,XTAU,ZERLAM,XLAM(1),XLAM(2)
      PRINT902
      PRINT903,XTAU,ZERLAM,XLAM(1),XLAM(2)
  902 FORMAT(1H123HPHASE I - TRUE SOLUTION /)
  903 FORMAT(1H0/
     1     1X11HTHICKNESS =, F10.4 /
     2     1X11HALBEDO(X) =, 20HA + B*TANH(10*(X-C))  //
     3  1X3HA =, E16.8, 10X3HB =, E16.8, 10X3HC =, E16.8 //)
      CALL NONLIN
      DO 3 I=1,N
      DO 3 J=1,N
    3 B2(I,J)=R2(I,J)
C
C
C             PHASE II
C
    4 READ1001,XTAU,ZERLAM,XLAM(1),XLAM(2)
      K=0
      PRINT904,K
      PRINT903,XTAU,ZERLAM,XLAM(1),XLAM(2)
C
      CALL NONLIN
C
  904 FORMAT(1H1   13HAPPROXIMATION, I3/ )
C
C             QUASILINEARIZATION ITERATIONS
```

```
C
      DO 5 K1=1,KMAX
      PRINT904,K1
      CALL PANDH
      CALL LINEAR
    5 CONTINUE
C
C
C
      READ1000,IGO
      GO TO (1,4),IGO
      END
$IBFTC DAUX
      SUBROUTINE DAUX
      DIMENSION V2(7,7),X(3),F(7),G(7)
      COMMON N,RT(7),WT(7),WR(7),AR(7,7),NPRNT,M1MAX,KMAX,DELTA,XTAU,
     1 ZERLAM,XLAM(2),B2(7,7),R2(7,7),IFLAG,R(28,101),T(1491),SIG,
     2 P(28,101),H(28,3,101),PTAU,PLAM(2),HTAU(3),HLAM(2,3),P2(7,7),
     3 H2(7,7,3),CONST(3),NEQ
      GO TO (1,2),IFLAG
C
CNONLINEAR
C
    1 L=3
      DO 4 I=1,N
      DO 4 J=1,I
      L=L+1
    4 V2(I,J)=T(L)
      DO 5 I=1,N
      DO 5 J=I,N
    5 V2(I,J)=V2(J,I)
      L=L+1
      VTAU=T(L)
      SIG=T(2)
      Y=VTAU*SIG
      X(1)=ZERLAM
      X(2)=XLAM(1)
      X(3)=XLAM(2)
      CALL ALBEDO(Y,X,Z)
      ZLAMDA=Z
C
      DO 6 I=1,N
      F(I)=0.0
      DO 7 K=1,N
    7 F(I)=F(I) + WR(K)*V2(I,K)
    6 F(I)=0.5*F(I) + 1.0
C
      DO 8 I=1,N
      DO 8 J=1,I
      L=L+1
      DR=-AR(I,J)*V2(I,J) + ZLAMDA*F(I)*F(J)
    8 T(L)=DR*VTAU
      DO 9 I=1,1
      L=L+1
    9 T(L)=0.0
```

```
      RETURN
C
C
CLINEAR
C
    2 SIG=T(2)
      Y=XTAU*SIG
      X(1)=ZERLAM
      X(2)=XLAM(1)
      X(3)=XLAM(2)
      CALL ALBEDO(Y,X,Z)
      ZLAMDA=Z
C
      DO 16 I=1,N
      F(I)=0.0
      DO 17 K=1,N
   17 F(I)=F(I) + WR(K)*R2(I,K)
   16 F(I)=0.5*F(I) + 1.0
C
CP'S
C
      L=3
      DO 14 I=1,N
      DO 14 J=1,I
      L=L+1
   14 V2(I,J)=T(L)
      DO 15 I=1,N
      DO 15 J=I,N
   15 V2(I,J)=V2(J,I)
      L=L+1
      VTAU=T(L)
C
      DO 10 I=1,N
      G(I)=0.0
      DO 10 K=1,N
   10 G(I)=G(I) + (V2(I,K)-R2(I,K))*WR(K)
      ARG=10.0*(Y-XLAM(2))
      PARTL=10.0*SIG*XLAM(1)*(1.0-(TANH(ARG))**2)
        M=3+NEQ
      DO 12 I=1,N
      DO 12 J=1,I
      FIJ=F(I)*F(J)
      CAPF=-AR(I,J)*R2(I,J) + ZLAMDA*FIJ
      T1=-XTAU*AR(I,J)*(V2(I,J)-R2(I,J))
      T2=XTAU*0.5*ZLAMDA*(F(I)*G(J)+F(J)*G(I))
      T3=XTAU*CAPF
      T4=(VTAU-XTAU)*(CAPF + XTAU*FIJ*PARTL)
        M=M+1
   12 T(M)=T1+T2+T3+T4
      DO 19 I=1,1
        M=M+1
   19 T(M)=0.0
C
CH'S
C
```

```
      DO 100 K=1,1
C
      DO 24 I=1,N
      DO 24 J=1,I
      L=L+1
   24 V2(I,J)=T(L)
      DO 25 I=1,N
      DO 25 J=I,N
   25 V2(I,J)=V2(J,I)
      L=L+1
      VTAU=T(L)
C
      DO 20 I=1,N
      G(I)=0.0
      DO 20 J=1,N
   20 G(I)=G(I) +  V2(I,J)*WR(J)
C
      DO 22 I=1,N
      DO 22 J=1,I
      FIJ=F(I)*F(J)
      CAPF=-AR(I,J)*R2(I,J) + ZLAMDA*FIJ
      T1=-XTAU*AR(I,J)*V2(I,J)
      T2=XTAU*0.5*ZLAMDA*(F(I)*G(J)+F(J)*G(I))
      T3=0.0
      T4=VTAU*(CAPF + XTAU*FIJ*PARTL)
        M=M+1
   22 T(M)=T1+T2+T3+T4
C
      DO 29 I=1,1
        M=M+1
   29 T(M)=0.0
  100 CONTINUE
      RETURN
      END
$IBFTC NONLIN
      SUBROUTINE NONLIN
      COMMON N,RT(7),WT(7),WR(7),AR(7,7),NPRNT,M1MAX,KMAX,DELTA,XTAU,
     1 ZERLAM,XLAM(2),B2(7,7),R2(7,7),IFLAG,R(28,101),T(1491),SIG,
     2 P(28,101),H(28,3,101),PTAU,PLAM(2),HTAU(3),HLAM(2,3),P2(7,7),
     3 H2(7,7,3),CONST(3),NEQ
C          NONLINEAR D.E. FOR TRUE SOLUTION OR FOR INITIAL APPROX.
C
      IFLAG=1
      T(2)=0.0
      T(3)=DELTA
      M=1
      L1=0
      L3=3
      DO 1 I=1,N
      DO 1 J=1,I
      L1=L1+1
      L3=L3+1
      R2(I,J)=0.0
      R(L1,M)=R2(I,J)
    1 T(L3)=R2(I,J)
```

```
      L3=L3+1
    2 T(L3)=XTAU
C
      NEQ=(N*(N+1))/2 + 1
      CALL INTS(T,NEQ,2,0,0,0,0,0,0)
C
      SIG=T(2)
      CALL OUTPUT
C
      DO 5 M1=1,M1MAX
      DO 4 M2=1,NPRNT
      CALL INTM
      M=M+1
      L1=0
      L3=3
      DO 3 I=1,N
      DO 3 J=1,I
      L1=L1+1
      L3=L3+1
      R2(I,J)=T(L3)
    3 R(L1,M)=R2(I,J)
    4 SIG=T(2)
    5 CALL OUTPUT
C
      RETURN
      END
$IBFTC LINEAR
      SUBROUTINE LINEAR
      DIMENSION CHKI(3)
      DIMENSION A(49,3),B(49),EMAT(50,50),          PIVOT(50),INDEX(50,2)
     1,IPIVOT(50),FVEC(50,1)
      COMMON N,RT(7),WT(7),WR(7),AR(7,7),NPRNT,M1MAX,KMAX,DELTA,XTAU,
     1  ZERLAM,XLAM(2),B2(7,7),R2(7,7),IFLAG,R(28,101),T(1491),SIG,
     2  P(28,101),H(28,3,101),PTAU,PLAM(2),HTAU(3),HLAM(2,3),P2(7,7),
     3  H2(7,7,3),CONST(3),NEQ
CBOUNDARY CONDITIONS
      MLAST=NPRNT*M1MAX + 1
      DO 1 K=1,1
      L=0
      DO 2 I=1,N
      DO 2 J=1,I
      L=L+1
    2 H2(I,J,K)=H(L,K,MLAST)
      DO 1 I=1,N
      DO 1 J=I,N
    1 H2(I,J,K)=H2(J,I,K)
      L=0
      DO 3 I=1,N
      DO 3 J=1,I
      L=L+1
    3 P2(I,J)=P(L,MLAST)
      DO 4 I=1,N
      DO 4 J=I,N
    4 P2(I,J)=P2(J,I)
CLEAST SQUARES
```

```
      DO 5 K=1,1
      L=0
      DO 5 I=1,N
      DO 5 J=1,N
      L=L+1
    5 A(L,K)=H2(I,J,K)
      L=0
      DO 6 I=1,N
      DO 6 J=1,N
      L=L+1
    6 B(L)=B2(I,J) - P2(I,J)
C
      LMAX=N**2
      PRINT60
   60 FORMAT(1H0)
      DO 61 L=1,LMAX
   61 PRINT82,(A(L,K),K=1,1),B(L)
C
      DO 8 I=1,1
      DO 7 J=1,1
      SUM=0.0
      DO 9 L=1,LMAX
    9 SUM=SUM + A(L,I)*A(L,J)
    7 EMAT(I,J)=SUM
      SUM=0.0
      DO 10 L=1,LMAX
   10 SUM=SUM + A(L,I)*B(L)
    8 FVEC(I,1)=SUM
C
      PRINT60
      DO 81 I=1,1
   81 PRINT82,(EMAT(I,J),J=1,1),FVEC(I,1)
   82 FORMAT(10X6E20.8)
C
      FVEC(1,1)=FVEC(1,1)/EMAT(1,1)
C
      DO 11 I=1,1
   11 CONST(I)=FVEC(I,1)
C
      XTAU    =CONST(1)
      PRINT903,XTAU,ZERLAM,XLAM(1),XLAM(2)
  903 FORMAT(1H0/
     1     1X11HTHICKNESS =, E16.8 /
     2     1X11HALBEDO(X) =, 20HA + B*TANH(10*(X-C))   //
     3  1X3HA =, E16.8, 10X3HB =, E16.8, 10X3HC =, E16.8 //)
C
CNEW APPROXIMATION
C
      M=1
      L=0
      DO 12 I=1,N
      DO 12 J=1,I
      L=L+1
      SUM=P(L,M)
      DO 13 K=1,1
```

```
   13   SUM =SUM + CONST(K)*H(L,K,M)
   12   R(L,M)=SUM
        L=0
        DO 14 I=1,N
        DO 14 J=1,I
        L=L+1
   14   R2(I,J)=R(L,M)
        SIG=0.0
        CALL OUTPUT
C
        DO 50 M1=1,M1MAX
        DO 18 M2=1,NPRNT
        M=M+1
        L=0
        DO 15 I=1,N
        DO 15 J=1,I
        L=L+1
        SUM=P(L,M)
        DO 16 K=1,1
   16   SUM=SUM + CONST(K)*H(L,K,M)
   15   R(L,M)=SUM
        L=0
        DO 17 I=1,N
        DO 17 J=1,I
        L=L+1
   17   R2(I,J)=R(L,M)
   18   SIG=SIG + DELTA
   50   CALL OUTPUT
C
        RETURN
        END
$IBFTC PANDH   LIST
        SUBROUTINE PANDH
        COMMON N,RT(7),WT(7),WR(7),AR(7,7),NPRNT,M1MAX,KMAX,DELTA,XTAU,
       1  ZERLAM,XLAM(2),B2(7,7),R2(7,7),IFLAG,R(28,101),T(1491),SIG,
       2  P(28,101),H(28,3,101),PTAU,PLAM(2),HTAU(3),HLAM(2,3),P2(7,7),
       3  H2(7,7,3),CONST(3),NEQ
        IFLAG=2
        T(2)=0.0
        T(3)=DELTA
        M=1
C P'S
C
        L1=0
        L3=3
        DO 1 I=1,N
        DO 1 J=1,I
        L1=L1+1
        L3=L3+1
        P(L1,M)=0.0
    1   T(L3)=P(L1,M)
        L3=L3+1
        PTAU=0.0
    2   T(L3)=PTAU
C
```

```
C H'S
C
      DO 7 K=1,1
      L1=0
      DO 3 I=1,N
      DO 3 J=1,I
      L1=L1+1
      L3=L3+1
      H(L1,K,M)=0.0
    3 T(L3)=H(L1,K,M)
C
      L3=L3+1
    6 HTAU(K)=1.0
    7 T(L3)=HTAU(K)
C
      L=0
      DO 8 I=1,N
      DO 8 J=1,I
      L=L+1
    8 R2(I,J)=R(L,M)
      DO 9 I=1,N
      DO 9 J=I,N
    9 R2(I,J)=R2(J,I)
C
      NEQ=2*((N*(N+1))/2 + 1)
      CALL INTS(T,NEQ,2,0,0,0,0,0,0)
      LMAX=(N*(N+1))/2
C
      DO 51 M1=1,M1MAX
      DO 50 M2=1,NPRNT
      CALL INTM
      M=M+1
CPREV.APPROX. R(I,J)
      L1=0
      DO 10 I=1,N
      DO 10 J=1,I
      L1=L1+1
   10 R2(I,J)=R(L1,M)
      DO 11 I=1,N
      DO 11 J=I,N
   11 R2(I,J)=R2(J,I)
      L1=0
      L3=3
      DO 12 I=1,N
      DO 12 J=1,I
      L1=L1+1
      L3=L3+1
   12 P(L1,M)=T(L3)
      L3=L3+1
      DO 13 K=1,1
      L1=0
      DO 14 I=1,N
      DO 14 J=1,I
      L1=L1+1
      L3=L3+1
```

```
   14   H(L1,K,M)=T(L3)
   13   L3=L3+1
   50   CONTINUE
   51   CONTINUE
        RETURN
        END
$IBFTC OUTPUT
        SUBROUTINE OUTPUT
        DIMENSION X(3)
        COMMON N,RT(7),WT(7),WR(7),AR(7,7),NPRNT,M1MAX,KMAX,DELTA,XTAU,
       1  ZERLAM,XLAM(2),B2(7,7),R2(7,7),IFLAG,R(28,101),T(1491),SIG,
       2  P(28,101),H(28,3,101),PTAU,PLAM(2),HTAU(3),HLAM(2,3),P2(7,7),
       3  H2(7,7,3),CONST(3),NEQ
        DO 1 I=1,N
        DO 1 J=I,N
    1   R2(I,J)=R2(J,I)
        Y=XTAU*SIG
        X(1)=ZERLAM
        X(2)=XLAM(1)
        X(3)=XLAM(2)
        CALL ALBEDO(Y,X,Z)
        PRINT100, SIG,Y,Z
  100   FORMAT(1H0 7HSIGMA =,F6.2, 4X5HTAU =, F6.2, 4X8HALBEDO =,F6.2/)
        DO 2 J=1,N
    2   PRINT101,J,(R2(I,J),I=1,N)
  101   FORMAT(I10, 7F10.6)
        RETURN
        END
$IBFTC ALBEDO
        SUBROUTINE ALBEDO(Y,X,Z)
        DIMENSION X(3)
        COMMON N,RT(7),WT(7),WR(7),AR(7,7),NPRNT,M1MAX,KMAX,DELTA,XTAU,
       1  ZERLAM,XLAM(2),B2(7,7),R2(7,7),IFLAG,R(28,101),T(1491),SIG,
       2  P(28,101),H(28,3,101),PTAU,PLAM(2),HTAU(3),HLAM(2,3),P2(7,7),
       3  H2(7,7,3),CONST(3),NEQ
        ARG=10.0*(Y-X(3))
        Z=X(1) + X(2)*TANH(ARG)
        RETURN
        END
```

Program B.3. Determination of the Two Albedos and the Thickness of the Lower Layer. The complete program consists of a MAIN program, and subroutines DAUX, NONLIN, PANDH, LINEAR, OUTPUT, and ALBEDO. The library routines MATINV and INTS/INTM are required.

```
$JOB            2609,STRAT3,HK0160,5,0,20,P
$PAUSE
$IBJOB STRAT2  MAP
$IBFTC RTINV
      COMMON N,RT(7),WT(7),WR(7),AR(7,7),NPRNT,M1MAX,KMAX,DELTA,XTAU,
     1  XLAM(3),        B2(7,7),R2(7,7),IFLAG,R(28,101),T(1491),SIG,
     2  P(28,101),H(28,3,101),PLAM(3),HLAM(3,3),P2(7,7),
     3  H2(7,7,3),CONST(3),NEQ
C
C            PHASE I
C
    1 READ1000,N
      PRINT899
      PRINT900,N
      READ1001,(RT(I),I=1,N)
      PRINT901,(RT(I),I=1,N)
      READ1001,(WT(I),I=1,N)
      PRINT901,(WT(I),I=1,N)
      DO 2 I=1,N
      WR(I)=WT(I)/RT(I)
      DO 2 J=1,N
    2 AR(I,J)= 1.0/RT(I) + 1.0/RT(J)
C
  899 FORMAT(1H146X36HRADIATIVE TRANSFER - INVERSE PROBLEM / )
 1000 FORMAT(6I12)
  900 FORMAT(6I20)
 1001 FORMAT(6E12.8)
  901 FORMAT(6E20.8)
      READ1000,NPRNT,M1MAX,KMAX
      PRINT900,NPRNT,M1MAX,KMAX
      READ1001,DELTA
      PRINT901,DELTA
      READ1001,XTAU,(XLAM(I),I=1,3)
      PRINT902
      PRINT903,XTAU,(XLAM(I),I=1,3)
  902 FORMAT(1H123HPHASE I - TRUE SOLUTION /)
  903 FORMAT(1H0/
     1     1X11HTHICKNESS =, F10.4 /
     2     1X11HALBEDO(X) =, 20HA + B*TANH(10*(X-C))  //
     3  1X3HA =, E16.8, 10X3HB =, E16.8, 10X3HC =, E16.8 //)
      CALL NONLIN
      DO 3 I=1,N
      DO 3 J=1,N
    3 B2(I,J)=R2(I,J)
C
C
C            PHASE II
C
    4 READ1001,XTAU,(XLAM(I),I=1,3)
      K=0
      PRINT904,K
      PRINT903,XTAU,(XLAM(I),I=1,3)
C
      CALL NONLIN
C
```

```
  904 FORMAT(1H1   13HAPPROXIMATION, I3/ )
C
C           QUASILINEARIZATION ITERATIONS
C
      DO 5 K1=1,KMAX
      PRINT904,K1
      CALL PANDH
      CALL LINEAR
    5 CONTINUE
C
C
C
      READ1000,IGO
      GO TO (1,4),IGO
      END
$IBFTC DAUX    LIST
      SUBROUTINE DAUX
      DIMENSION V2(7,7),X(3),F(7),G(7)
     1  ,VLAM(3)
      COMMON N,RT(7),WT(7),WR(7),AR(7,7),NPRNT,M1MAX,KMAX,DELTA,XTAU,
     1  XLAM(3),        B2(7,7),R2(7,7),IFLAG,R(28,101),T(1491),SIG,
     2  P(28,101),H(28,3,101),PLAM(3),HLAM(3,3),P2(7,7),
     3  H2(7,7,3),CONST(3),NEQ
      GO TO (1,2),IFLAG
C
CNONLINEAR
C
    1 L=3
      DO 4 I=1,N
      DO 4 J=1,I
      L=L+1
    4 V2(I,J)=T(L)
      DO 5 I=1,N
      DO 5 J=I,N
    5 V2(I,J)=V2(J,I)
      DO 51 I=1,3
      L=L+1
   51 VLAM(I)=T(L)
      SIG=T(2)
      Y=XTAU*SIG
      DO 52 I=1,3
   52 X(I)=VLAM(I)
      CALL ALBEDO(Y,X,Z)
      ZLAMDA=Z
C
      DO 6 I=1,N
      F(I)=0.0
      DO 7 K=1,N
    7 F(I)=F(I) + WR(K)*V2(I,K)
    6 F(I)=0.5*F(I) + 1.0
C
      DO 8 I=1,N
      DO 8 J=1,I
      L=L+1
      DR=-AR(I,J)*V2(I,J) + ZLAMDA*F(I)*F(J)
```

```
    8   T(L)=DR
        DO 9 I=1,3
        L=L+1
    9   T(L)=0.0
        RETURN
C
C
CLINEAR
C
    2   SIG=T(2)
        Y=XTAU*SIG
        DO 21 I=1,3
   21   X(I)=XLAM(I)
        CALL ALBEDO(Y,X,Z)
        ZLAMDA=Z
C
        DO 16 I=1,N
        F(I)=0.0
        DO 17 K=1,N
   17   F(I)=F(I) + WR(K)*R2(I,K)
   16   F(I)=0.5*F(I) + 1.0
C
CP'S
C
        L=3
        DO 14 I=1,N
        DO 14 J=1,I
        L=L+1
   14   V2(I,J)=T(L)
        DO 15 I=1,N
        DO 15 J=I,N
   15   V2(I,J)=V2(J,I)
        DO 18 I=1,3
        L=L+1
   18   VLAM(I)=T(L)
C
        DO 10 I=1,N
        G(I)=0.0
        DO 10 K=1,N
   10   G(I)=G(I) + (V2(I,K)-R2(I,K))*WR(K)
        ARG=10.0*(Y-XLAM(3))
        TARG=TANH(ARG)
        XTANX=-10.0*XLAM(2)*(1.0-TARG**2)
          M=3+NEQ
        DO 12 I=1,N
        DO 12 J=1,I
        FIJ=F(I)*F(J)
        CAPF=-AR(I,J)*R2(I,J) + ZLAMDA*FIJ
        T1=CAPF
        T2=-AR(I,J)*(V2(I,J)-R2(I,J))
       1  + 0.5*ZLAMDA*(F(I)*G(J) + F(J)*G(I))
        T3=(VLAM(1)-XLAM(1))*FIJ
        T4=(VLAM(2)-XLAM(2))*TARG*FIJ
        T5=(VLAM(3)-XLAM(3))*XTANX*FIJ
          M=M+1
```

```
  12   T(M)=T1+T2+T3+T4+T5
       DO 19 I=1,3
         M=M+1
  19   T(M)=0.0
C
CH'S
C
       DO 100 K=1,3
C
       DO 24 I=1,N
       DO 24 J=1,I
       L=L+1
  24   V2(I,J)=T(L)
       DO 25 I=1,N
       DO 25 J=I,N
  25   V2(I,J)=V2(J,I)
       DO 26 I=1,3
       L=L+1
  26   VLAM(I)=T(L)
C
       DO 20 I=1,N
       G(I)=0.0
       DO 20 J=1,N
  20   G(I)=G(I) +  V2(I,J)*WR(J)
C
       DO 22 I=1,N
       DO 22 J=1,I
       FIJ=F(I)*F(J)
       T1=0.0
       T2=-AR(I,J)*V2(I,J) + 0.5*ZLAMDA*(F(I)*G(J) + F(J)*G(I))
       T3=VLAM(1)*FIJ
       T4=VLAM(2)*TARG*FIJ
       T5=VLAM(3)*XTANX*FIJ
         M=M+1
  22   T(M)=T1+T2+T3+T4+T5
C
       DO 29 I=1,3
         M=M+1
  29   T(M)=0.0
 100   CONTINUE
       RETURN
       END
$IBFTC NONLIN
       SUBROUTINE NONLIN
       COMMON N,RT(7),WT(7),WR(7),AR(7,7),NPRNT,M1MAX,KMAX,DELTA,XTAU,
      1  XLAM(3),        B2(7,7),R2(7,7),IFLAG,R(28,101),T(1491),SIG,
      2  P(28,101),H(28,3,101),PLAM(3),HLAM(3,3),P2(7,7),
      3  H2(7,7,3),CONST(3),NEQ
C            NONLINEAR D.E. FOR TRUE SOLUTION OR FOR INITIAL APPROX.
C
       IFLAG=1
       T(2)=0.0
       T(3)=DELTA
       M=1
       L1=0
```

```
      L3=3
      DO 1 I=1,N
      DO 1 J=1,I
      L1=L1+1
      L3=L3+1
      R2(I,J)=0.0
      R(L1,M)=R2(I,J)
    1 T(L3)=R2(I,J)
      DO 2 I=1,3
      L3=L3+1
    2 T(L3)=XLAM(I)
C
      NEQ=(N*(N+1))/2 + 3
      CALL INTS(T,NEQ,2,0,0,0,0,0,0)
C
      SIG=T(2)
      CALL OUTPUT
C
      DO 5 M1=1,M1MAX
      DO 4 M2=1,NPRNT
      CALL INTM
      M=M+1
      L1=0
      L3=3
      DO 3 I=1,N
      DO 3 J=1,I
      L1=L1+1
      L3=L3+1
      R2(I,J)=T(L3)
    3 R(L1,M)=R2(I,J)
    4 SIG=T(2)
    5 CALL OUTPUT
C
      RETURN
      END
$IBFTC PANDH
      SUBROUTINE PANDH
      COMMON N,RT(7),WT(7),WR(7),AR(7,7),NPRNT,M1MAX,KMAX,DELTA,XTAU,
     1 XLAM(3),          B2(7,7),R2(7,7),IFLAG,R(28,101),T(1491),SIG,
     2 P(28,101),H(28,3,101),PLAM(3),HLAM(3,3),P2(7,7),
     3 H2(7,7,3),CONST(3),NEQ
      IFLAG=2
      T(2)=0.0
      T(3)=DELTA
      M=1
C P'S
C
      L1=0
      L3=3
      DO 1 I=1,N
      DO 1 J=1,I
      L1=L1+1
      L3=L3+1
      P(L1,M)=0.0
    1 T(L3)=P(L1,M)
```

```
      DO 2 I=1,3
      L3=L3+1
      PLAM(I)=0.0
    2 T(L3)=PLAM(I)
C
C H'S
C
      DO 7 K=1,3
      L1=0
      DO 3 I=1,N
      DO 3 J=1,I
      L1=L1+1
      L3=L3+1
      H(L1,K,M)=0.0
    3 T(L3)=H(L1,K,M)
C
      DO 7 I=1,3
      L3=L3+1
      HLAM(I,K)=0.0
      IF(I-K)7,6,7
    6 HLAM(I,K)=1.0
    7 T(L3)=HLAM(I,K)
C
      L=0
      DO 8 I=1,N
      DO 8 J=1,I
      L=L+1
    8 R2(I,J)=R(L,M)
      DO 9 I=1,N
      DO 9 J=I,N
    9 R2(I,J)=R2(J,I)
C
      NEQ=4*((N*(N+1))/2 + 3)
      CALL INTS(T,NEQ,2,0,0,0,0,0,0)
      LMAX=(N*(N+1))/2
      PRINT52,T(2),(P(L,M),H(L,1,M),L=1,LMAX)
   52 FORMAT(1H0F9.4,5E20.8/(10X5E20.8))
C
      DO 51 M1=1,M1MAX
      DO 50 M2=1,NPRNT
      CALL INTM
      M=M+1
CPREV.APPROX. R(I,J)
      L1=0
      DO 10 I=1,N
      DO 10 J=1,I
      L1=L1+1
   10 R2(I,J)=R(L1,M)
      DO 11 I=1,N
      DO 11 J=I,N
   11 R2(I,J)=R2(J,I)
      L1=0
      L3=3
      DO 12 I=1,N
      DO 12 J=1,I
```

```
      L1=L1+1
      L3=L3+1
   12 P(L1,M)=T(L3)
      L3=L3+3
      DO 13 K=1,3
      L1=0
      DO 14 I=1,N
      DO 14 J=1,I
      L1=L1+1
      L3=L3+1
   14 H(L1,K,M)=T(L3)
   13 L3=L3+3
   50 CONTINUE
   51 PRINT52,T(2),(P(L,M),H(L,1,M),L=1,LMAX)
      RETURN
      END
$IBFTC LINEAR
      SUBROUTINE LINEAR
      DIMENSION CHKI(3)
      DIMENSION A(49,3),B(49),EMAT(50,50),          PIVOT(50),INDEX(50,2)
     1,IPIVOT(50),FVEC(50,1)
      COMMON N,RT(7),WT(7),WR(7),AR(7,7),NPRNT,M1MAX,KMAX,DELTA,XTAU,
     1  XLAM(3),        B2(7,7),R2(7,7),IFLAG,R(28,101),T(1491),SIG,
     2  P(28,101),H(28,3,101),PLAM(3),HLAM(3,3),P2(7,7),
     3  H2(7,7,3),CONST(3),NEQ
CBOUNDARY CONDITIONS
      MLAST=NPRNT*M1MAX + 1
      DO 1 K=1,3
      L=0
      DO 2 I=1,N
      DO 2 J=1,I
      L=L+1
    2 H2(I,J,K)=H(L,K,MLAST)
      DO 1 I=1,N
      DO 1 J=I,N
    1 H2(I,J,K)=H2(J,I,K)
      L=0
      DO 3 I=1,N
      DO 3 J=1,I
      L=L+1
    3 P2(I,J)=P(L,MLAST)
      DO 4 I=1,N
      DO 4 J=I,N
    4 P2(I,J)=P2(J,I)
CLEAST SQUARES
      DO 5 K=1,3
      L=0
      DO 5 I=1,N
      DO 5 J=1,N
      L=L+1
    5 A(L,K)=H2(I,J,K)
      L=0
      DO 6 I=1,N
      DO 6 J=1,N
      L=L+1
```

```
    6   B(L)=B2(I,J) - P2(I,J)
C
        LMAX=N**2
        PRINT60
   60   FORMAT(1H0)
        DO 61 L=1,LMAX
   61   PRINT82,(A(L,K),K=1,3),B(L)
C
        DO 8 I=1,3
        DO 7 J=1,3
        SUM=0.0
        DO 9 L=1,LMAX
    9   SUM=SUM + A(L,I)*A(L,J)
    7   EMAT(I,J)=SUM
        SUM=0.0
        DO 10 L=1,LMAX
   10   SUM=SUM + A(L,I)*B(L)
    8   FVEC(I,1)=SUM
C
        PRINT60
        DO 81 I=1,3
   81   PRINT82,(EMAT(I,J),J=1,3),FVEC(I,1)
   82   FORMAT(10X6E20.8)
C
        CALL MATINV(EMAT,3,FVEC,1,DETERM,PIVOT,INDEX,IPIVOT)
C
        DO 11 I=1,3
   11   CONST(I)=FVEC(I,1)
C
        DO 20 I=1,3
   20   XLAM(I)=CONST(I)
        PRINT903,XTAU,(XLAM(I),I=1,3)
  903   FORMAT(1H0/
     1  1X11HTHICKNESS =, E16.8 /
     2     1X11HALBEDO(X) =, 20HA + B*TANH(10*(X-C))  //
     3  1X3HA =, E16.8, 10X3HB =, E16.8, 10X3HC =, E16.8 //)
C
CNEW APPROXIMATION
C
        M=1
        L=0
        DO 12 I=1,N
        DO 12 J=1,I
        L=L+1
        SUM=P(L,M)
        DO 13 K=1,3
   13   SUM =SUM + CONST(K)*H(L,K,M)
   12   R(L,M)=SUM
        L=0
        DO 14 I=1,N
        DO 14 J=1,I
        L=L+1
   14   R2(I,J)=R(L,M)
        SIG=0.0
        CALL OUTPUT
```

```
C
      DO 50 M1=1,M1MAX
      DO 18 M2=1,NPRNT
      M=M+1
      L=0
      DO 15 I=1,N
      DO 15 J=1,I
      L=L+1
      SUM=P(L,M)
      DO 16 K=1,3
   16 SUM=SUM + CONST(K)*H(L,K,M)
   15 R(L,M)=SUM
      L=0
      DO 17 I=1,N
      DO 17 J=1,I
      L=L+1
   17 R2(I,J)=R(L,M)
   18 SIG=SIG + DELTA
   50 CALL OUTPUT
C
      RETURN
      END
$IBFTC OUTPUT
      SUBROUTINE OUTPUT
      DIMENSION X(3)
      COMMON N,RT(7),WT(7),WR(7),AR(7,7),NPRNT,M1MAX,KMAX,DELTA,XTAU,
     1  XLAM(3),          B2(7,7),R2(7,7),IFLAG,R(28,101),T(1491),SIG,
     2  P(28,101),H(28,3,101),PLAM(3),HLAM(3,3),P2(7,7),
     3  H2(7,7,3),CONST(3),NEQ
      DO 1 I=1,N
      DO 1 J=I,N
    1 R2(I,J)=R2(J,I)
      Y=XTAU*SIG
      DO 3 I=1,3
    3 X(I)=XLAM(I)
      CALL ALBEDO(Y,X,Z)
      PRINT100, SIG,Y,Z
  100 FORMAT(1H0 7HSIGMA =,F6.2, 4X5HTAU =, F6.2, 4X8HALBEDO =,F6.2/)
      DO 2 J=1,N
    2 PRINT101,J,(R2(I,J),I=1,N)
  101 FORMAT(I10, 7F10.6)
      RETURN
      END
$IBFTC ALBEDO
      SUBROUTINE ALBEDO(Y,X,Z)
      DIMENSION X(3)
      COMMON N,RT(7),WT(7),WR(7),AR(7,7),NPRNT,M1MAX,KMAX,DELTA,XTAU,
     1  XLAM(3),          B2(7,7),R2(7,7),IFLAG,R(28,101),T(1491),SIG,
     2  P(28,101),H(28,3,101),PLAM(3),HLAM(3,3),P2(7,7),
     3  H2(7,7,3),CONST(3),NEQ
      ARG=10.0*(Y-X(3))
      Z=X(1) + X(2)*TANH(ARG)
      RETURN
      END
$ENTRY          RTINV
          7
25446046E-0112923441E-0029707742E-0050000000E 0070292258E 0087076559E 00
97455396E 00
64742484E-0113985269E-0019091502E-0020897958E-0019091502E-0013985269E-00
64742484E-01
        10          10          4
      0.01
       1.0          0.5         0.1         0.5
       1.0          0.6         .09         0.4
$IBSYS
```

APPENDIX C

LIBRARY ROUTINES

C-1 Library Routine INTS/INTM

This is documented in

R. Causey, W. L. Frank, W. L. Sibley, and F. Valadez, "RS W031 - RW INT, Adams-Moulton, Runge-Kutta Integration IBMAP Coded Subroutine (FORTRAN IV)", RAND 7044 Library Routine W031, The Rand Corporation, Santa Monica, 1964.

This is a subroutine for the numerical integration of ordinary differential equations. Let NEQ be the number of dependent variables. The one dimensional array T is in COMMON. The contents of the T region are as follows.

T(2)	Current value of independent variable
T(3)	Step size of integration
T(4)...T(3+NEQ)	Dependent variables
T(4+NEQ)...T(3+2*NEQ)	Derivatives

Initial set-up for integration consists of entering initial values of independent variable and all NEQ dependent variables, followed by the statement

```
CALL INTS(T,NEQ,N1,0,0,0,0,0,0),
```

where N1 = 2 for Adams-Moulton integration method.

To take one integration step forward, the statement required is

```
CALL INTM.
```

This results in the new values of independent and dependent variables being entered into their locations, and current values of the respective derivatives entered as above.

This routine requires that the differential equations be coded in subroutine DAUX, and the independent variable, dependent variables, and derivatives are to be referenced by their locations in the T array. It must be understood that the derivative of T(3) is to be entered into T(4+NEQ), and so forth.

C-2 Library Routine MATINV

This routine is documented in

S. Belcher and B. S. Barbow, "RS W019 - ANF402 - MATRIX Inversion with Accompanying Solution of Linear Equations", RAND 7044 Library Routine W019, The Rand Corporation, Santa Monica, 1964.

This subroutine solves for x in the equation,

$$A\ x = b,$$

where A is an $N \times N$ matrix, and b and x are column vectors of length N. The calling statement is

```
CALL MATINV (A, N, B, 1, D, P, I, IP),
```

where A is the matrix, B is the right-hand vector, D is the determinant, and P, I and IP are other arrays not utilized in the listed programs. The solution vector is returned in the array B, whose original contents are destroyed.

INDEX

Absorption coefficient 105
Adams-Moulton method 18,24,26
Adaptive control 189
Agranovich, Z. S. 28,122
Albedo 31,63,78
Ambarzumian, V. A. 58,143
Anisotropic scattering 83

Bellman, R. 28,29,30,33, 57,58,59,122,143, 163,178,196,197, 199,220,229,242
Blore, W. E. 163
Boltyanskii, V. G. 220
Borg, G. 143
Boundary value problems 12,108,202,203
Brillouin, L. 143
Bucker, H. 178
Bucy, R. 196,199,220
Buell, J. 29,178,229
Busbridge, I. 59

Cannon, J. R. 28
Cardiogram 232
Cardiology 231
Casti, J. 29
Chandrasekhar, S. 33, 40,58,122
Chen, J. A. 98
Chu, C. M. 98
Churchill, S. W. 98
Clasen, R. J. 82
Collier, C. 241
Convergence 17,24,70, 176-177
Cost 201
Courant, R. 143,178
Cox, H. 196
Criterion 75

Dantzig, G. 82
Dave, J. V. 98
Deirmendjian, D. 98
Detchmendy, D. M. 29,178, 199,220
Differential-difference equations 221
Differential equations, ordinary 32,40-41, 146
Diffusion equation 124
Dipole, dipole moment 231
Dreyfus, S. E. 122
Dubyago, A. D. 30
Dynamic programming 104

Eykhoff, P. 28

Fermat's principle 167
Filtering 179
Finite difference equations 162
Fourier series 88,162
Frank, P. 58,143
Fujita, H. 59

Gamkrelidze, R. V. 220
Gaussian noise 73
Gaussian quadrature 40-42, 126
Gradient method 171
Grant, F. 178
Greenstein, J. L. 99

Henrici, P. 30

Henyey, L. G. 99
Hilbert, D. 143,178
Hille, E. 58
Ho, Y. 196
Ho, Y. C. 28
Homogeneous solution 14, 20-23
Horak, H. G. 84,98
Huss, R. 29
Hyodo, T. 59

Index of refraction 145, 166
Initial value problems 3,18,109
Integral-differential equations 32,39, 63,87-88
Intensity of radiation 35,38
Internal intensities 103
Interpolation 199
Invariant imbedding 32-39, 85,149-152,181,203
Inverse problems 1,31,62
Irvine, W. M. 98

Jacobian matrix 14
Jacquez, J. 229

Kagiwada, H. H. 28,29,30,33, 58,59,98,122,143,178, 196,220,242
Kalaba, R. E. 28,29,30,33,57, 58,59,98,122,143,163, 178,196,220,229,242
Kalman, R. 196,199,220
Kobayashi, K. 59
Kyhl, R. L. 163

Laplace
  transforms 125-127
  transforms, numerical inversion of 127,131
Leacock, J. A. 98
Lee, R.C.K. 28
Legendre functions
  associated 89
  shifted 40-42
Legendre polynomials 127
Lenoble, J. 98
Lindsay, R. B. 163
Linear algebraic equations 15,18,23,69
Linear differential equations 14-15,19
Linear programming 77
Linear two-point boundary value problems 139
Lyttleton, R. A. 30

Malkevich, M. S. 99
Marchenko, V. A. 28,122
Maslennikov, M. V. 28
Melsa, J. L. 28
Middleton, D. 196
Mikhlin, S. 178
Miller, R. J. 28
Milne, W. E. 30
Minimization 45,180,202
Minnaert, M. 59
Mischenko, E. F. 220
Models 77
Morris, H. 178
Morse, P. M. 163
Multiple scattering 31,61, 83

Neutron transport 101
Newton's method 16,173
Noisy observations 61, 73-75,191
Nonlinear boundary value problems 3,11,12, 48,63,127,168,222
Nonlinear differential equations 179,190
Nonlinear equations 119
Nonlinear filtering 179,199
Numerical integration 18,23

Optical thickness 31,43

Optimal estimate of state 183,201
Optimality, principle of 105
Osterberg, H. 163

Page, W. A. 28
Partial differential equations 2,124,181,204
Particular solution 14,20-23
Pearson, R. 241
Pedersen, M. 178
Phase function 83-84,88
Phillips, R. 58
Pontryagin, L. S. 220
Potential 7,18
Potentials, skin 231
Preisendorfer, R. W. 28,99
Prestrud, M. C. 29,30,57,59, 143,196
Primich, R. I. 163
Pugachev, V. 196

Quadratically convergent method 172
Quasilinearization 11-18, 49,66,92,128,157, 168-171,224,236
Radiative transfer 31,61,83
Rays 165
Reflection coefficient 107, 148
Reflection functions 34,84, 85
Reflected wave 147
Riccati equation 39,40,151
Robillard, P. E. 163
Runge-Kutta method 18

Sage, A. P. 28
Schelkunoff, S. A. 163
Scher, A. 231,241
Schilling, G. F. 98
Scott, M. R. 58
Search procedure 171
Sekera, Z. 98
Selvester, R. 241
Shaules, A. 99
Shimizu, A. 59
Sims, A. R. 28,122
Smolitskiy, K. 178
Sobolev, V. V. 33,58, 59,98
Spingarn, K. 28
Sridhar, R. 29,30,199,220
Successive approximations 12,25-26,53,56,72, 136,161
Sutton, R. E. 28
Swerling, P. 196
System identification problems 1,11,189
System parameters 11

Time lags 221
Tompkins, C. B. 178
Transmission coefficient 111
Transmitted wave 147
Transport theory 101
Trapezoidal rule of integration 18
Tyler, J. E. 99

Ueno, S. 59,220

Van de Hulst, H. C. 98
Von Moses, R. 58,143

Wave equation 125,145
Wave number 145
Wave propagation 145
Wernick, A. 99
West, G. 178
White, D. 178
Wiener, N. 196
Wing, G. M. 29,58,59,196

Yang, C. 29
Yound, A. 241